WITHDRAWN

The Conjugacy Problem and Higman Embeddings

WITHDRAWN

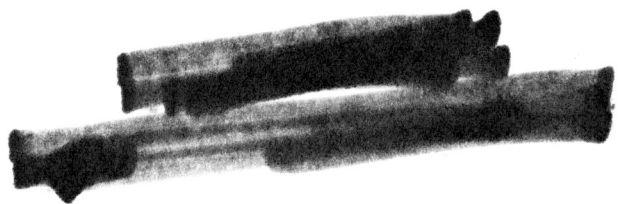

Memoirs
of the
American Mathematical Society

Number 804

The Conjugacy Problem and Higman Embeddings

A. Yu. Ol'shanskii
M. V. Sapir

July 2004 • Volume 170 • Number 804 (first of 4 numbers) • ISSN 0065-9266

American Mathematical Society
Providence, Rhode Island

2000 *Mathematics Subject Classification.*
Primary 20E07, 20F06, 20F10.

Library of Congress Cataloging-in-Publication Data

Ol'shanskii, A. IU. (Aleksandr IUr'evich)
The conjugacy problem and Higman embeddings/A. Yu. Ol'shanskii, M.V. Sapir.
   p. cm. — (Memoirs of the American Mathematical Society, ISSN 0065-9266 ; no. 804)
"Volume 170, number 804 (first of 4 numbers)."
Includes bibliographical references.
ISBN 0-8218-3513-0 (alk. paper)
   1. Frattini subgroups. 2. Conjugacy classes. 3. Embeddings (Mathematics) I. Sapir, Mark, 1957– II. Title. III. Series.
QA3.A57 no. 804
[QA177]
510 s–dc22
[512′.23]
                                                                                                                           2004046105

Memoirs of the American Mathematical Society

    This journal is devoted entirely to research in pure and applied mathematics.

    Subscription information. The 2004 subscription begins with volume 167 and consists of six mailings, each containing one or more numbers. Subscription prices for 2004 are $583 list, $466 institutional member. A late charge of 10% of the subscription price will be imposed on orders received from nonmembers after January 1 of the subscription year. Subscribers outside the United States and India must pay a postage surcharge of $31; subscribers in India must pay a postage surcharge of $43. Expedited delivery to destinations in North America $35; elsewhere $130. Each number may be ordered separately; *please specify number* when ordering an individual number. For prices and titles of recently released numbers, see the New Publications sections of the *Notices of the American Mathematical Society*.

    Back number information. For back issues see the *AMS Catalog of Publications*.

    Subscriptions and orders should be addressed to the American Mathematical Society, P. O. Box 845904, Boston, MA 02284-5904, USA. *All orders must be accompanied by payment.* Other correspondence should be addressed to 201 Charles Street, Providence, RI 02904-2294, USA.

    Copying and reprinting. Individual readers of this publication, and nonprofit libraries acting for them, are permitted to make fair use of the material, such as to copy a chapter for use in teaching or research. Permission is granted to quote brief passages from this publication in reviews, provided the customary acknowledgment of the source is given.

    Republication, systematic copying, or multiple reproduction of any material in this publication is permitted only under license from the American Mathematical Society. Requests for such permission should be addressed to the Acquisitions Department, American Mathematical Society, 201 Charles Street, Providence, Rhode Island 02904-2294, USA. Requests can also be made by e-mail to reprint-permission@ams.org.

    Memoirs of the American Mathematical Society is published bimonthly (each volume consisting usually of more than one number) by the American Mathematical Society at 201 Charles Street, Providence, RI 02904-2294, USA. Periodicals postage paid at Providence, RI. Postmaster: Send address changes to Memoirs, American Mathematical Society, 201 Charles Street, Providence, RI 02904-2294, USA.

    © 2004 by the American Mathematical Society. All rights reserved.
This publication is indexed in *Science Citation Index*®, *SciSearch*®, *Research Alert*®, *CompuMath Citation Index*®, *Current Contents*®/*Physical, Chemical & Earth Sciences*.
Printed in the United States of America.

    ∞ The paper used in this book is acid-free and falls within the guidelines established to ensure permanence and durability.
Visit the AMS home page at http://www.ams.org/

    10 9 8 7 6 5 4 3 2 1   09 08 07 06 05 04

Contents

1. **Introduction** — 1
2. **List of relations** — 10
   - 2.1 S-machines — 10
   - 2.2 An auxiliary group — 11
   - 2.3 The hardware of $\mathcal{S}$ — 11
   - 2.4 Notation — 13
   - 2.5 The rules of the S-machine $\mathcal{S}$ — 14
   - 2.6 The S-machines $\bar{\mathcal{S}}$ and $\mathcal{S} \cup \bar{\mathcal{S}}$ — 17
   - 2.7 Converting $\mathcal{S}$ into a set of relations — 20
   - 2.8 Converting $\bar{\mathcal{S}}$ into a set of relations — 22
   - 2.9 The presentation of the group $\mathcal{H}$ — 22
3. **The first properties of $\mathcal{H}$** — 22
   - 3.1 Diagrams — 22
   - 3.2 $\mathcal{G}$ embeds into $\mathcal{H}$ — 23
   - 3.3 Bands and annuli — 27
   - 3.4 Forbidden annuli — 32
   - 3.5 Surgeries involving $\mathcal{G}$-cells — 34
   - 3.6 Shifting indexes — 35
4. **The group $\mathcal{H}_2$** — 37
5. **The word problem in $\mathcal{H}_1$** — 42
6. **Some special diagrams** — 48
   - 6.1 θ-bands and trapezia — 48
   - 6.2 Trapezia with 2-letter bases — 52
   - 6.3 Trapezia simulate the work of S-machines — 53
7. **Computations of $\mathcal{S} \cup \bar{\mathcal{S}}$** — 60
   - 7.1 Brief history — 62
   - 7.2 Standard computations — 64
   - 7.3 Tame and wild computations — 73
   - 7.4 Computations of $\bar{\mathcal{S}}$ — 78
   - 7.5 All ring computations — 83
8. **Spirals** — 86
9. **Rolls** — 97
   - 9.1 Rolls without a base — 98
   - 9.2 Rolls with L-annuli — 101
   - 9.3 Rolls with base $L_{j-1}^{-1} K_j L_j$ — 109

9.4 Arbitrary rolls 118

10 Arrangement of hubs   **122**

11 The end of the proof   **127**

   References   **128**

   Subject index   **131**

Abstract

For every finitely generated recursively presented group $\mathcal{G}$ we construct a finitely presented group $\mathcal{H}$ containing $\mathcal{G}$ such that $\mathcal{G}$ is (Frattini) embedded into $\mathcal{H}$ and the group $\mathcal{H}$ has solvable conjugacy problem if and only if $\mathcal{G}$ has solvable conjugacy problem. Moreover $\mathcal{G}$ and $\mathcal{H}$ have the same r.e. Turing degrees of the conjugacy problem. This solves a problem by D. Collins.

Key words and phrases: Conjugacy problem, Higman embedding, finitely presented group, recursively presented group, Turing degree

2000 Mathematics Subject Classification: 20E07, 20F06, 20F10

1 Introduction

In 1961, G. Higman [Hi] published the celebrated theorem that a finitely generated group is recursively presented if and only if it is a subgroup of a finitely presented group. Along with the results of Novikov [Nov] and Boone [Bo] this result showed that objects from logic (in that case, recursively enumerable sets) have group theoretic characterizations (see Manin [Ma] for the philosophy of Higman embeddings).

Clapham [Cla] (see also corrections in [Va]) was probably the first to investigate properties preserved under Higman embeddings. In particular, he slightly modified the original Higman construction and showed that his embedding preserves solvability (and even the r.e. Turing degree) of the word problem. Valiev [Va] sketched a proof of a much stronger result: every finitely generated recursively presented group G embeds in a finitely presented group H such that the word problem in H polynomially reduces to the word problem in G. In particular, if the word problem of a finitely generated group is in P (i.e. can be solved in polynomial time by a deterministic Turing machine) then it can be embedded into a finitely presented group whose word problem is also in P (a correction to Valiev's paper was published in Mathematical reviews, review 54 # 413, see also Lyndon and Schupp [LS] and Manin [Ma]). A simplified version of Valiev's construction (but not a proof of the polynomial reducibility) was later published by Aanderaa and Cohen in [AC] (see also Kalorkoti [Kal]). In [BORS], Birget, Rips and the authors of this paper obtained a group theoretic characterization of NP (non-deterministic polynomial time): a finitely generated group G has word problem in NP if and only if G is embedded into a finitely presented group with polynomial Dehn function.

The conjugacy problem turned out to be much harder to preserve under embeddings. Collins and Miller [CM] and Gorjaga and Kirkinskiĭ[GK] proved that even subgroups of index 2 of finitely presented groups do not inherit solvability or unsolvability of the conjugacy problem.

In 1976 D. Collins [KT] posed the following question (problem 5.22): *Does there exist a version of the Higman embedding theorem in which the degree of unsolvability of the conjugacy problem is preserved?*

Received by the editor November 21, 2002.

Both authors were supported in part by the NSF grant DMS 0072307. In addition, the research of the first author was supported in part by the Russian Fund for Basic Research 99-01-00894 and by the INTAS grant 99-1224, the research of the second author was supported in part by the NSF grant DMS 9978802 and the US-Israeli BSF grant 1999298.

It was quickly realized that the main problem would be in preserving the smallest Turing degree, that is the solvability of conjugacy problem. In 1980, [Col], Collins analyzed existing proofs of Higman's theorem, and discovered that there are essential difficulties. If a finitely generated group C is embedded into a finitely presented group B using any existing at that time constructions then the solvability of conjugacy problem in B implies certain properties of C which are much stronger than solvability of conjugacy problem. Collins even wrote the following pessimistic comments: "There seems at present to be no hope to establishing the analogue of Clapham's theorem. ... Furthermore these difficulties seem to be more or less inevitable given the structure of the proof and probably a wholly new strategy will be needed to avoid them. For the present the most one can be hoped for is the isolation of conditions on C that are necessary and sufficient for the preservation of the solvability of the conjugacy problem in the Higman embedding."

In this paper, we do find a "new strategy" and prove the following result. Recall (see R. Thompson [Tho]) that a subgroup $\mathcal{G}$ of a group $\mathcal{H}$ is called *Frattini embedded* if any two elements of $\mathcal{G}$ that are conjugate in $\mathcal{H}$ are also conjugate in $\mathcal{G}$. Clearly if $\mathcal{G}$ is Frattini embedded into $\mathcal{H}$ and $\mathcal{H}$ has solvable conjugacy problem then $\mathcal{G}$ has solvable conjugacy problem (by results from [CM] and [GK], cited above non-Frattini embedded subgroups do not necessarily inherit solvability of the conjugacy problem).

Theorem 1.1. *A finitely generated group has solvable conjugacy problem if and only if it is Frattini embedded into a finitely presented group with solvable conjugacy problem.*

Moreover we prove the following stronger result solving the original problem of Collins:

Theorem 1.2. *A finitely generated group $\mathcal{G}$ with recursively enumerable set of relations has conjugacy problem of r.e. Turing degree α if and only if $\mathcal{G}$ can be Frattini embedded into a finitely presented group $\mathcal{H}$ with conjugacy problem of r.e. Turing degree α.*

In the forthcoming paper [OlSa4], we will present some corollaries of these theorems (we did not include the proofs of them here in order not to increase the difficulty level unnecessarily). In particular, we will show that one can drop the restriction that $\mathcal{G}$ is finitely generated in Theorems 1.1, 1.2, replacing it with "countably generated". We shall also show that a finitely generated group with solvable word, power and order problems can be embedded into a finitely presented group with solvable word, order, power, and conjugacy problems. Thus a Higman embedding can greatly improve algorithmic properties of a group.

In this section, we explain the main ideas which lead us to our construction. We try to keep notation as simple as possible in this section of the paper, so the notation here does not necessarily coincide with the notation in the rest of the paper.

We consider the case when $\mathcal{G} = \langle A \mid \mathcal{E} \rangle$ has decidable conjugacy problem and we want to embed it into a finitely presented group with decidable conjugacy problem (the case when the conjugacy problem in $\mathcal{G}$ has arbitrary r.e. Turing degree is similar). The standard idea is to start with a machine which recognizes the set $\mathcal{E}$ and then interpret this machine in a group. Of course we would like to have a machine which is easier to interpret. The most suitable machines for our purposes are the so called S-machines invented by the second author in [SBR] and successfully used in [SBR, BORS, OlSa1, OlSa2, OlSa3]. An S-machine works with words which can have complicated structure determined by the problem. Different parts of these words can be elements of different groups (so we do not distinguish words whose corresponding parts are equal in the corresponding groups). As in Turing machines, every rule of an S-machine replaces subwords containing state letters by other subwords. For an exact definition of S-machines see Section 2.1. For now the reader can imagine just an ordinary Turing machine which works with words from the free group instead of a free monoid.

Here we give just one example of an S-machine which essentially goes back to C. Miller [Mil] (although Miller did not use S-machines). The real S-machine used in this paper is a "descendant" of this S-machine (obtained from Miller's machine by several "mutations"). Assume that A is a symmetric set that is it contains the formal inverses of its elements, and let us embed $\mathcal{G} = \langle A \mid \mathcal{E} \rangle$ into *some* finitely presented group $\bar{\mathcal{G}} = \langle A \cup Y \mid \bar{\mathcal{E}} \rangle$ using, say, the standard Higman embedding. Now consider the S-machine $\mathcal{M}$ with one state letter q and the tape alphabet $A \cup Y$ considered as a generating set of the free group, and the following commands (each command says which subwords of a word should be replaced and the replacement word):

- $[q \to a^{-1}qa]$, for every $a \in A \cup Y$,

- $[q \to rq]$, for every $r \in \bar{\mathcal{E}}$.

The first set of rules allows the state letter to move freely along a word of the form uqv where u, v are words from the free group on $A \cup Y$. Such words are called *admissible words* for $\mathcal{M}$.

The second set of rules allows us to insert any relation from $\bar{\mathcal{E}}$ into the word (we are also allowed to insert and remove subwords of the form aa^{-1} because $A \cup Y$ is a generating set of a free group).

It is easy to see that a word qu, where u is a word over A, is recognized by this machine (that is the machine takes it to the word q) if and only if $u = 1$ in the group $\bar{\mathcal{G}}$, that is if and only $u = 1$ in $\mathcal{G}$ (because $\mathcal{G}$ is embedded into $\bar{\mathcal{G}}$).[1]

---

[1] Notice that by a result of Sapir [OlSa2], for every Turing machine T there exists an

Now let us present the ideas of a Higman embedding based on $\mathcal{M}$. Let $T_1, T_2, ..., T_m = W_0$ be an accepting *computation* of $\mathcal{M}$, that is a sequence of admissible words such that T_{i+1} is obtained from T_i by applying an S-rule of $\mathcal{M}$, the last word in this sequence is equal to q in the free group.

Suppose that modulo some set of group relations Z we have for every $i = 1, ..., m-1$

$$P_i T_i = T_{i+1} P_i'$$

for some words P_i and P_i' over a certain alphabet (it is included in the finite generating set of the group we are constructing), that is we have a van Kampen diagram with the boundary label $P_i T_i (P_i')^{-1} T_{i+1}^{-1}$ (Figure 1):

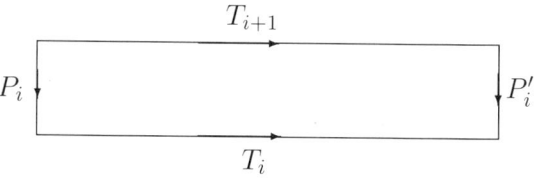

Figure 1.

(the set Z describes the method of tessellating this rectangle into cells). Then we have that $PT_1 = W_0 P'$ where $P = P_{m-1}...P_2 P_1$, $P_i' = P_{m-1}'...P_2' P_1'$. The corresponding van Kampen diagram Δ has the form of a *trapezium*:

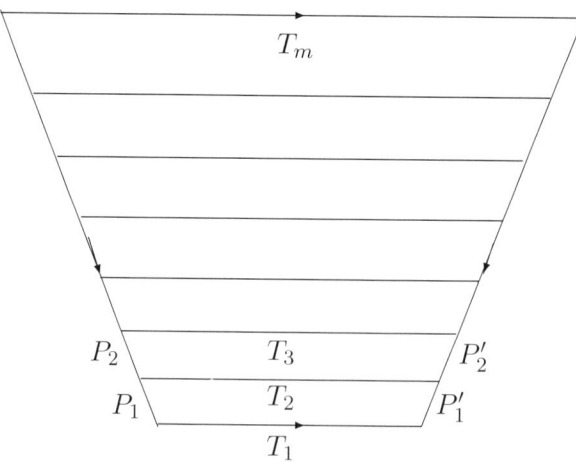

Figure 2.

For example, if we use the S-machine $\mathcal{M}$ described above then the presentation Z is easy to describe: for every rule $\tau = [q \to uqv]$ we have the

---

S-machine $\mathcal{M}$ corresponding a to a group $\bar{\mathcal{G}}$ as above which is polynomially equivalent to T. This shows that machines $\mathcal{M}$ are as powerful as ordinary Turing machines.

following relations in Z: $\theta_\tau^{-1} q \theta_\tau = uqv$, $\theta_\tau a = a\theta_\tau$ for every $a \in A \cup Y$ (the letters θ_τ, corresponding to the rules of $\mathcal{M}$ are added to the generating set). In this case the words P_i, P_i' are of length 1, and are equal to θ_τ.

The words P and P' contain the *history* of our computation, because P_i and P_i' correspond to S-rules applied during the computation. The words T_1 and T_m are labels of the bottom and the top paths of the trapezium.

If the set of relations Z is chosen carefully (and there are very many ways of doing this, see [SBR, BORS, OlSa2, OlSa1, OlSa3]), then the converse is also true:

Condition 1. If there exists a trapezium with the top labelled by T_1, the bottom labelled by W_0 and the sides labelled by words over Θ (the set of all θ_τ) then T_1 is admissible and the labels of the sides correspond to the history of an accepting computation.

Thus the word $PT_1(P')^{-1}W_0^{-1}$ can be written on the boundary of a trapezium if and only if the word T_1 is accepted.

The next step is to consider $N \geq 1$ copies of our trapezium with labels taken from different alphabets. For technical reasons (similar to the hyperbolic graphs argument in [SBR], [Ol2], [BORS]) we need N to be even and large enough (say, $N \geq 8$). Let us choose disjoint alphabets in different copies of the trapezium. Then we can glue two copies of the trapezium by using a special letter k which conjugates letters from the right side of the first trapezium with letters from the left side of the mirror image of another copy (Figure 3):

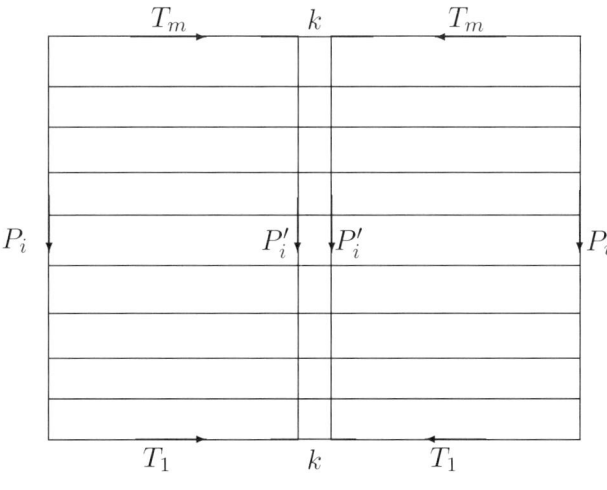

Figure 3.

(Notice that if P and P' are copies of each other then we can also glue the right side of one trapezium to the left side of another without taking mirror images.) Of course the conjugacy relations involving letter k should be added into the set of group relations Z.

Suppose that T_m is equal to q, i.e. the word T_1 is *accepted* by the machine. If we connect the first copy of Δ with the second copy, then with the third copy, and finally connect the N-th copy with the first copy, using different letters $k_2, ..., k_N, k_1$, we get an *annular* diagram, a *ring*. The outer boundary of this ring has label $\mathcal{K}(T_1)$ of the form $k_1 T_1' k_2^{-1} (T_1'')^{-1}...$ (here we use the fact that N is even). The words $T_1', ..., T_1^{(N)}$ are different copies of T_1. The inner boundary has label $k_1 q' k_2^{-1} (q'')^{-1}...$ (this word is called the *hub*) (Figure 4 shows the diagram in the case $N = 4$):

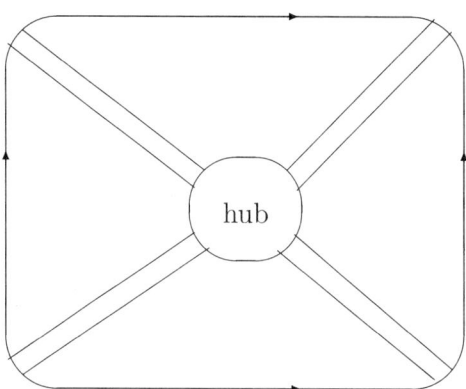

Figure 4.

We add the hub to the list of defining relations Z. Now with every word T of the form uqv where u, v are words in $A \cup Y$ we have associated the word $\mathcal{K}(T)$, and we see that if T is accepted then $\mathcal{K}(T)$ is equal to 1 modulo Z. Again, if Z is chosen carefully then the converse is also true:

Condition 2. If $\mathcal{K}(T)$ is equal to 1 modulo Z then T is accepted by $\mathcal{S}$. Thus we have an interpretation of $\mathcal{S}$ in the group generated by the tape alphabet and the state alphabet of $\mathcal{S}$, the set Θ, and the set $\{k_1, ..., k_N\}$, subject to the relations from Z.

The diagram on the previous Figure is called a *disk*. The trapezia forming this disk are numbered in the natural order (from 1 to N).

We assume that the copy of the alphabet A used in the first subtrapezium of a disk coincides with A itself (the generating set of $\mathcal{G}$).

Let us denote the group given by the presentation Z we have got so far by $G(\mathcal{M})$.

There are several ways to use the group $G(\mathcal{M})$ to embed $\mathcal{G}$ into a finitely presented group. We cannot use the method used by Higman [Hi], Aanderaa [AC], Rotman [Rot], and us in [BORS] because it leads to problems discovered by Collins in [Col]. We use another method, described in [OlSa2] and used also in [OlSa3].

Consider $N-1$ new copies of the trapezium Δ. Number these trapezia by $2', ..., N'$. Identify the copy of A (but not Y) and the state letter q in the trapezium number i' with the copy of A and the copy of q in the "old" trapezium number i. Now glue the trapezium number $2'$ with the trapezium number $3'$ using letter k_3 (as before), then glue in the trapezium number $4'$ and so on, but glue the trapezium number N' with the trapezium number $2'$ using $k_1 q k_2^{-1}$ (that would be possible because as it will turn out in that case the words P, P' will be copies of each other). That is conjugation of the letters on the side of the trapezium number N' by $k_1 q k_2^{-1}$ gives the letters of the side of the mirror image of the trapezium number $2'$. Let us add the relations used in the new trapezia and the conjugation relations (involving k_i) to Z.

The resulting picture is an annular diagram. The inner boundary of it is labelled by the hub. If we add the hub to the diagram, we get a van Kampen diagram which also looks like a disk but has N-1 sectors. The outer boundary of this new disk is labelled by the word $k_1 q k_2^{-1} V$ where V is the suffix of $\mathcal{K}(qu)$ staying after k_2^{-1}. Thus the word $k_1 q k_2^{-1} V$ and the word $k_1 q u k_2^{-1} V$ are both equal to 1 modulo Z, so Z implies $u = 1$. Let us call the group given by the set of relations Z that we have constructed by $\mathcal{H}$.

We have proved that the identity map on A can be extended to a homomorphism from $\mathcal{G}$ to the subgroup $\langle A \rangle$ of the group $\mathcal{H}$ given by the set of relations Z. It is possible to show that this homomorphism is injective, so $\mathcal{G}$ is embedded into $\mathcal{H}$.

Suppose that $\mathcal{G}$ has solvable conjugacy problem. Is it true that $\mathcal{H}$ has solvable conjugacy problem? The answer is "not always". Let us give two examples.

Example 1. Consider two pairs of words u_1, u_2 and v_1, v_2 over the alphabet A (we can view these words as elements of $\mathcal{G}$). Let q be the first state letter in $\mathcal{K}(qu)$. Consider the words $W_1 = q^{-1} u_1 q u_2 q^{-1}$ and $W_2 = q^{-1} v_1 q v_2 q^{-1}$. Suppose that there exists a word $W(A)$ in the alphabet A such that $W(A) u_1 W(A)^{-1} = v_1$, $W(A) u_2 W(A)^{-1} = v_2$. Let $W(A) = a_1 a_2 ... a_k$, $a_i \in A$. For every $a \in A$ let $\theta(a)$ correspond to the S-rule of the form $[q \to a^{-1} q a]$. Consider the word $W(\Theta) = \theta(a_1) \theta(a_2) ... \theta(a_k)$. Then it is easy to see that $W(\theta) W_1 W(\theta^{-1}) = W_2$. Thus if the pair (u_1, u_2) is conjugated to the pair (v_1, v_2) in $\mathcal{G}$ then the words W_1 and W_2 are conjugate in $\mathcal{G}$. One can also prove that the converse statement is also true. In this example, pairs of words can be obviously replaced by any t-tuples of words, $t \geq 2$. Therefore if the conjugacy problem is solvable in $\mathcal{G}$ then the conjugacy of t-tuples of elements in $\mathcal{G}$ is solvable. It is known [Col] that there exists a finitely generated group $\mathcal{G}$ with solvable conjugacy problem and unsolvable problem of conjugacy for sequences of elements. For such a group $\mathcal{G}$ the group $\mathcal{H}$ has undecidable conjugacy problem.

Example 2. Take two pairs of words $(u_1, u_2), (v_1, v_2)$ over one of the copies of the alphabet A, say, $A' \neq A$, let q' be the corresponding copy of the letter q. Then it is possible to show that the words $q'u_1(q')^{-1}u_2q'$ and $q'v_1(q')^{-1}v_2q'$ (where u_1, u_2, v_1, v_2 are words in A') are conjugate in $\mathcal{H}$ if and only if for some words P, Q

$$Qu_1Q^{-1} = v_1, P^{-1}u_2P = v_2$$

in the free group and $PQ = 1$ in $\mathcal{G}(A')$, a copy of $\mathcal{G}$. This problem easily reduces (exercise!) to the following problem about group $\mathcal{G}$: *given four words s, t, p, r in the alphabet A find two integers m, n such that $st^n = p^m r$ in $\mathcal{G}$.* It is quite possible that this problem can be undecidable even if the conjugacy problem is decidable.

These two examples suggest that we need to change the machine $\mathcal{M}$: we cannot allow trapezia with top path labelled by words of the form $qu_1q^{-1}u_2q$ having long histories. In other words, if a computation of the S-machine starts with $qu_1q^{-1}u_2q$, then it should not be too long (more precisely, if it is too long, we should be able to replace it with a shorter computation). If we achieve that then we would have a recursive bound on the length of the word W in Example 1 and the words P, Q in Example 2, so we would be able to find these words by a simple search.

This lead us to the S-machine $\mathcal{S}$ considered in this paper (see Sections 2.3, 2.4, 2.5). This machine has four sets of state letters: K, L, P, R-letters. These letters are placed in an admissible word $\mathcal{K}(u)$ according to the following pattern: $KLPRK^{-1}R^{-1}P^{-1}L^{-1}KLPR....$

The P-letters play the role of the many copies of q in $\mathcal{K}(qu)$ above: they move along an admissible word and "find" places where to insert relations from $\bar{\mathcal{E}}$. The K-letters are the letters k_i in $\mathcal{K}(qu)$ (they divide admissible words into copies of the same word and do not move). The L- and R-letters are "support letters". While P is moving, L, R stay next to K-letters. But when P "wants" to insert a relation, L, R must move next to P and they insert the relation "together" (executing the rule $LPR \to rLPR$, $r \in \bar{\mathcal{E}}$). After that L, R must move back to their initial positions, and P can move to a new place in the admissible word (the work of the machine resembles a complicated dance performed synchronously by several groups of dancers).

It is important to mention that we cannot allow all parts of the admissible words to be arbitrary group words. To avoid long computations without inserting new relation from $\bar{\mathcal{E}}$, we had to require that the words between neighbor K-letter and L-letter, L-letter and P-letter, R-letter and K-letter (but not between P-letter and R-letter!) to be positive (i.e. they cannot contain letters a^{-1}).

In order to keep these subwords positive we use a trick that goes back to Novikov, Boone and Higman (see Rotman [Rot]). They used a special letter x and Baumslag-Solitar relations of the form $a^{-1}xa = x^2$ for every a in the tape alphabet of the machine. The idea is that if we have relations $a^{-1}xa = x^2$ for all $a \in A$ and a word $u^{-1}xu$, where u is a reduced word in

A, is equal modulo these relations to a power of x then the first letter of u is positive.

We use a similar idea. But our task is more complicated than in [Nov, Bo, Hi] because we need to consider conjugacy of arbitrary pairs of words, not only those that have "nice" structure. So we had to use many different letters x, replace 2 by 4 in the Baumslag-Solitar relations, and make some other technical modifications. Notice that one problem with using the x-letters and Baumslag-Solitar relations is that since $N - 1$ is odd we cannot use x-letters in the second disk described above (indeed, otherwise it would be impossible to glue the $N - 1$ sectors together since the words P and P' would not be copies of each other). Thus we need to consider two S-machines (one responsible for the first disk, another for the second disk) which are very similar but have different "hardware": the admissible words of the first machine must have positive parts described above, the admissible words of the second machine do not have to satisfy this restriction.

Notice that Examples 1 and 2 are only the main obstacles that we had to overcome in this paper. Other technical difficulties lead to further fine tuning of our S-machine. Sections 2.7, 2.8, 2.9 contain precise description of the presentation of our group $\mathcal{H}$.

Now let us briefly describe the strategy of the proof that the conjugacy problem in $\mathcal{H}$ is Turing reducible to the conjugacy problem in $\mathcal{G}$. As in our previous papers the main tool in this paper is bands (other people call them "strips", "corridors", etc.).

In terms of annular (Schupp) diagrams, our task is the following: given two words u and v in generators of $\mathcal{H}$, find out if there exists an annular diagram over the presentation of $\mathcal{H}$ with boundary labels u and v. It turns out that any annular diagram over $\mathcal{H}$ (after removing some parts of recursively bounded size) becomes a diagram of one of three main types: a *ring* (where the boundary labels contain K-, L-, P-, R-letters but do not contain θ-letters, all maximal θ-bands are annuli surrounding the hole of the diagram), a *roll* (where the boundary label does not contain K-, L-, P-, R-letters but contain θ-letters; all maximal K-, L-, P-, R-bands are annuli surrounding the hole), and a *spiral* (where the boundary labels contain both K-, L-, P-, or R-letters and θ-letters; both K-, L-, P-, R-bands and the θ-bands start and end on the different boundary components, and each θ-band crosses each K-, L-, P-, and R-band many times[2]). Different methods are used to treat different cases. Roughly speaking, the study of rings amounts to study the lengths of computations of our S-machines, the study of rolls amounts to the study of the space complexity (how much space is needed by the machines during a computation), and the study of spirals amounts to the study of computations with periodic history. The x-letters and the

---

[2] In that case the θ-band looks like a spiral on a plane that starts at the origin and crosses certain straight lines (the K-, L-, P- and R-bands) many times.

Baumslag-Solitar relations allow us to treat the case of rings, but they cause the main technical difficulties in the cases of rolls and spirals.

Acknowledgment. The authors are grateful to the referee for many useful remarks.

2 List of relations

2.1 S-machines

Here we give the general definition of S-machines (see [OlSa2]). The precise definition of the S-machines used in this paper will be given later.

Let $T_1, ..., T_k$ be a collection of groups, $T_i = \langle A_i \rangle$, $i = 1, ..., k$, and the sets A_i of generators pairwise disjoint. Let $\mathcal{K}$ be a set of *state letters*, disjoint from $\cup A_i$. Let **A** be a set of *admissible* words of the form $q_1 w_1 q_2 ... q_k w_k q_{k+1}$ where w_i is a word over some A_j, $q_i \in \mathcal{K} \cup \mathcal{K}^{-1}$.

We do not distinguish words over A_i which are equal in the group T_i. So we can view w_i as elements in T_i.

Any S-*rule* τ has the following form $[k_1 \to v_1 k'_1 u_1, ..., k_n \to v_n k'_n u_n]$ where u_j, v_j are group words over $\cup A_i$, $k_j \in \mathcal{K} \cup \mathcal{K}^{-1}$. Every S-rule is a partial transformation on the set of admissible words. Every admissible word W in the domain of the S-rule τ must have $\mathcal{K}$-letters only from the set $\{k_1^{\pm 1}, ..., k_n^{\pm 1}\}$ (not all of these letters may occur, and some of them may occur more than once). This transformation takes an admissible word W, and replaces each $k_j^{\pm 1}$ by $(v_j k'_j u_j)^{\pm 1}$, freely reduces the resulting word and then trims maximal prefix and suffix of the resulting word consisting of non-$\mathcal{K}$-letters. For example, if $W \equiv k_1 p_1 k_1 p_2 k_2^{-1} p_3 k_3$ for some words p_1, p_2, p_3 over $\cup A_i$ (we assume $n \geq 3$) is in the domain of τ, then the result of the application of τ is the freely reduced form of $k'_1 u_1 p_1 v_1 k'_1 u_1 p_2 u_2^{-1} k_2^{-1} v_2^{-1} p_3 v_3 k'_3$. Notice that the resulting word must be an admissible word for the S-machine as well, otherwise W is not in the domain of τ.

The *inverse* rule τ^{-1} is the inverse partial transformation. It has the form $[k'_1 \to v_1^{-1} k_1 u_1^{-1}, ..., k'_n \to v_n^{-1} k_n u_n^{-1}]$.

Thus we can view an S-machines as a semigroup of partial transformations of the set of admissible words. If W is an admissible word of an S-machine $\mathcal{S}$, and $h = \tau_1 \tau_2 ... \tau_n$ is a sequence (word) of rules of $\mathcal{S}$, applicable to W (that is W is in the domain of the partial map h), then the result W' of application of h to W is denoted by $W \circ h$. The corresponding computation $W, W \circ \tau_1, W \circ \tau_1 \tau_2, ..., W'$ will be denoted by $W \bullet h$ or $W \bullet h = W'$.

The collection of groups T_i and the set of admissible words form the *hardware* of an S-machine. The *software* of an S-machine is a finite collection of S-rules closed under taking inverses.

2.2 An auxiliary group

Let $\mathcal{G}$ be a group given by the following presentation

$$\mathcal{G} = \langle a_1, \ldots, a_m | w = 1, w \in \mathcal{E} \rangle, \tag{2.1}$$

where $\mathcal{E}$ is a recursively enumerable set of words in $a_1, \ldots, a_m$. Adding if necessary, new generators and relations of the form $a_i a'_i = 1$ one can assume that the set $\mathcal{E}$ consists of positive words only, i.e., there are no occurrences of a_i^{-1} in the words $w \in \mathcal{E}$.

We first prove Theorem 1.1 and then show how to modify the proof to obtain Theorem 1.2. So we shall assume that $\mathcal{G}$ has solvable conjugacy problem.

By the Higman Embedding Theorem [Rot] there is a finitely presented group $\bar{\mathcal{G}}$ generated by $a_1, \ldots, a_m, \ldots, a_{\bar{m}}$ containing $\mathcal{G}$ as a subgroup. By Clapham-Valiev's result [Cla, Va], one can assume that the word problem in $\bar{\mathcal{G}}$ is decidable.

Besides, one may assume that again, for each generator a of $\bar{\mathcal{G}}$, the inverse letter a' is also included in the set of generators $\{a_1, \ldots, a_m, \ldots, a_{\bar{m}}\}$ of $\bar{\mathcal{G}}$. For every $a \in \{a_1, \ldots, a_{\bar{m}}\}$, we assume that there are positive relators of the form aa' and $a'a$ in the finite set $\bar{\mathcal{E}}$ of positive defining relators for $\bar{\mathcal{G}}$. We will also suppose that if $r \in \bar{\mathcal{E}}$ then $r' \in \bar{\mathcal{E}}$ where r' is obtained from r^{-1} by replacing every occurrence of a letter a^{-1} by the letter a'. Finally, we will assume that $\bar{\mathcal{E}}$ contains the empty word $\emptyset$. It is clear that $\bar{\mathcal{E}}$ is a presentation of the group $\bar{\mathcal{G}}$ in the class of all monoids, so we have the following

Lemma 2.1. *Assume that $w_1 = w_2$ for positive group words w_1, w_2 in the generators of $\bar{\mathcal{G}}$. Then there is a sequence of positive words starting with w_1 and ending with w_2, where every word in the sequence is obtained from the previous word by insertion or deletion of a subword $w \in \bar{\mathcal{E}}$.*

The list of relations of the group $\mathcal{H}$ which we are going to construct will depend on the set $\bar{\mathcal{E}}$ of defining words for $\bar{\mathcal{G}}$. Lemma 3.9 below claims that there is a natural embedding of the subgroup $\mathcal{G} \leq \bar{\mathcal{G}}$ into $\mathcal{H}$, but (caution!) we will not embed the whole group $\bar{\mathcal{G}}$ into $\mathcal{H}$.

2.3 The hardware of $\mathcal{S}$

The group $\mathcal{H}$ that we are going to construct is associated with two very similar S-machines, $\mathcal{S}$ and $\bar{\mathcal{S}}$. We shall describe the S-machines, then we will describe how to convert these machines into a group presentation.

We fix an even number $N \geq 8$.

The set $\mathcal{K}$ of state letters of $\mathcal{S}$ consists of letters $z_j(r, i)$ where $z \in \{K, L, P, R\}$, $j = 1, \ldots, N$, $r \in \bar{\mathcal{E}}$, $i \in \{1, 2, 3, 4, 5\}$.

We also define the set of *basic* letters $\tilde{\mathcal{K}}$ which consists of letters z_j where $z \in \{K, L, P, R\}$, $j = 1, ..., N$. Letters from $\tilde{\mathcal{K}}^{-1}$ are also called *basic* letters. There exists a natural map from $\mathcal{K} \cup \mathcal{K}^{-1}$ to $\tilde{\mathcal{K}} \cup \tilde{\mathcal{K}}^{-1}$ which forgets indexes r and i. If z is a basic letter, $r \in \bar{\mathcal{E}}$, $i \in \{1, 2, 3, 4, 5\}$ then $z(r, i) \in \mathcal{K}$. If U is a word in $\tilde{\mathcal{K}}$ and other letters, $r \in \bar{\mathcal{E}}$, $i \in \{1, 2, 3, 4, 5\}$ then $U(r, i)$ is a word obtained from U by replacing every letter $z \in \tilde{\mathcal{K}}$ by $z(r, i)$. The parameters r and i in the letter $z(r, i)$ or in the word $U(r, i)$ will be called the $\bar{\mathcal{E}}$-coordinate and the Ω-coordinate of the word.

The set $\mathcal{A}$ of *tape letters* of the machine $\mathcal{S}$ consists of letters $a_i(z)$ where $i = 1, ..., \bar{m}$, $z \in \tilde{\mathcal{K}}$. For every $z \in \tilde{\mathcal{K}}$ we define $\mathcal{A}(z)$ as the set of all $a_i(z)$, $i = 1, ..., \bar{m}$.

Let $\tilde{\Sigma}$ be the following word (considered as a cyclic word):

$$K_1 L_1 P_1 R_1 K_2^{-1} R_2^{-1} P_2^{-1} L_2^{-1} K_3 L_3 P_3 R_3 K_4^{-1} \\ ... K_N^{-1} R_N^{-1} P_N^{-1} L_N^{-1} \qquad (2.2)$$

Notice that for every basic letter z precisely one of z and z^{-1} occurs in the word $\tilde{\Sigma}$. The word $\Sigma = \tilde{\Sigma}(\emptyset, 1)$ will be called the *hub*.

For every $z \in \tilde{\mathcal{K}} \cup \tilde{\mathcal{K}}^{-1}$ by z_- we denote the letter immediately preceding z in the cyclic word $\tilde{\Sigma}$ or in $\tilde{\Sigma}^{-1}$. This definition is correct because every basic letter occurs exactly once in the word $\tilde{\Sigma}$ or its inverse (see remark above). If $z' = z_-$ then we set $z = z'_+$. Similarly we define z_- and z_+ for $z \in \mathcal{K}$. Notice that for every $j = 1, ..., N$,

$$(L_j)_- = \begin{cases} K_j & \text{if } j \text{ is odd,} \\ K_{j+1}^{-1} & \text{if } j \text{ is even.} \end{cases}$$

and

$$(R_j)_+ = \begin{cases} K_{j+1}^{-1} & \text{if } j \text{ is odd,} \\ K_j & \text{if } j \text{ is even.} \end{cases}$$

To simplify the notation and avoid extra parentheses, we shall denote $(L_j)_-$ by $\overleftarrow{L}_j$ and $(R_j)_+$ by $\overrightarrow{R}_j$. We also define $\mathcal{A}(z^{-1})$ for $z \in \tilde{\mathcal{K}}$ by setting $\mathcal{A}(z^{-1}) = \mathcal{A}(z_-)$.

The language of *admissible words* of the machine $\mathcal{S}$ consists of all reduced words of the form $W \equiv y_1 u_1 y_2 u_2 ... y_t u_t y_{t+1}$ where $y_1, ..., y_{t+1} \in \mathcal{K}^{\pm 1}$, u_i are words in $\mathcal{A}(y_i)$, $i = 1, 2, ..., t$, and for every $i = 1, 2, ..., t$, either $y_{i+1} \equiv (y_i)_+$ or $y_{i+1} \equiv y_i^{-1}$. (Here $\equiv$ is the letter-for-letter, or graphical, equality of words.) The projection of $y_1...y_{t+1}$ onto $\tilde{\mathcal{K}}$ will be called the *base* of the admissible word W. The subword $y_i u_i y_{i+1}$ is called the $y_i y_{i+1}$-*sector* of the admissible word W, $i = 1, 2.....$ The word u_i is called the *inner part* of the $y_i y_{i+1}$-sector. We assume that the inner part of any zL_j-, zP_j- or $R_j z$-sector of W and W^{-1} is a positive word ($z \in \tilde{\mathcal{K}}$), that is it does not contain a^{-1} for any $a \in \mathcal{A}$. Notice that if W is an admissible word for $\mathcal{S}$ then W^{-1} is an admissible word as well.

By definition, all $\mathcal{K}$-letters in an admissible word have the same $\bar{\mathcal{E}}$-coordinates and Ω-coordinates. Notice that $\tilde{\Sigma}(r,i)$ is an admissible word for every r,i.

Two admissible words W, W' are considered *equal* if $W \equiv W'$.

2.4 Notation

In this paper, the S-rules will have a specific form which allows us to simplify the notation. In every rule $[k_1 \to v_1 k_1' u_1, ..., k_n \to v_n k_n' u_n]$, the projections k_i and k_i' on $\tilde{\mathcal{K}}$ will be always the same, the $\bar{\mathcal{E}}$-coordinates and the Ω-coordinates of all state letters $k_1, ..., k_n$ (resp. $k_1', ..., k_n'$) will be the same. In addition, each u_i (resp. v_i) will contain only letters from $\mathcal{A}(k_i)$ (resp. $\mathcal{A}((k_i)_-)$), $i = 1, ..., n$.

For every word W in the alphabet $\{a_1, ..., a_{\bar{m}}\}$ and every $k \in \tilde{\mathcal{K}}$, $W(k)$ will denote the word obtained from W by substitution $a_i \to a_i(k)$, $i = 1, ..., \bar{m}$.

In any expressions like vku where v is a word in $\mathcal{A}(k_-)$, u is a word in $\mathcal{A}(k)$, $k \in \tilde{\mathcal{K}}$, we shall not mention what alphabets these words are written in. For example the notation $a_i P_j a_i^{-1}$ means $a_i(L_j) P_j a_i(P_j)^{-1}$, that is the two a_i are brothers taken from different alphabets.

Thus in order to simplify notation, we shall write a rule τ of the S-machines S in the form

$$[k_1 \to v_1 k_1 u_1, ..., k_n \to v_n k_n u_n; r \to r', \omega \to \omega']$$

where $k_i \in \tilde{\mathcal{K}}$, v_i are words in $\mathcal{A}((k_i)_-)$, u_i are words in $\mathcal{A}(k_i)$. We shall denote v_i by $v(\tau((k_i)_-))$, u_i by $u_\tau(k_i)$.

Some of the arrows in a rule τ can have the form $k_i \overset{\ell}{\to}$. This means that the rule τ can be applied to an admissible word W only if for every $k_i k_{i'}$-sector and every $k_{i''}(k_i)_+$-sector in W, its inner part is the empty word, and $k_{i'} = (k_i)_+$, $k_{i''} = k_i$ (i.e. $k_{i'} \neq k_i^{-1}$ and $k_{i''} \neq (k_i)_+^{-1}$). We shall say that in this case the $k_i(k_i)_+$-sectors are *locked* by the rule and the rule *locks* these sectors. If kk_+-sectors are locked by the rule, $u(k)$ and $v(k)$ must be empty. Thus if τ locks kk_+-sectors then τ^{-1} locks these sectors as well.

Let us expand the definition of $u_\tau(k), v_\tau(k_-)$ to the base letters not occurring in the rule by setting that in this case $u_\tau(k)$ and $v_\tau(k_-)$ are empty.

Thus with every rule τ and every $k \in \mathcal{K}$ we associate two words over $\mathcal{A}$: $u_\tau(k)$ and $v_\tau(k)$. We expand this definition to $k \in \tilde{\mathcal{K}}^{-1}$ by saying that $v_\tau(k^{-1}) \equiv u_\tau(k)^{-1}$, $u_\tau(k^{-1}) \equiv v_\tau(k_-)^{-1}$.

By definition, to *apply* the rule $\tau = [z_1 \to v_1 z_1 u_1, ..., z_s \to v_s z_s u_s; r \to r', \omega \to \omega']$ to an admissible word $W = y_1 w_1 y_2 ... w_k y_{k+1}$ means replacing each $y_i(r, \omega)$ by $v_i y_i(r', \omega') u_i$, reducing the resulting word, and trimming the prefix $v_\tau((y_1)_-)$ from the beginning and the suffix $u_\tau(y_{k+1})$ from the end of the resulting word. (Notice that w_i is non-empty if $y_i \equiv y_{i+1}^{-1}$, and therefore $y_i(r', \omega')$ will not cancel with $y_{i+1}(r', \omega')$ in the resulting word, because the subword between these two letters will not be empty.) The rule τ is called *applicable* to W if:

(1) for every $z \in \tilde{\mathcal{K}}$ such that τ locks zz_+-sectors the inner parts of zz_+-sectors in W are empty words, and W does not have zz^{-1}-sectors or $z_+^{-1}z_+$-sectors.

(2) the resulting word W' is again admissible (that is the inner parts of all sectors which are supposed to be negative or positive are such).

2.5 The rules of the S-machine S

The rules from $S^+ \subset S$, described below, will be called *positive*, the inverses of these rules will be called *negative*, each rule of S will be either positive or negative.

The set S^+ consists of ten subsets $S^+(\omega)$, where ω is one of the symbols $1, 12, 2, 23, 3, 34, 4, 45, 5, 51$.

The set $S^+(1)$ consists of rules $\tau(1, \emptyset, i)$, where $i = 1, ..., \bar{m}$. The rule $\tau(1, \emptyset, i)$ has the form

$$[\overleftarrow{L}_j \overset{\ell}{\to} \overleftarrow{L}_j, P_j \to a_i P_j a_i^{-1}, R_j \overset{\ell}{\to} R_j, j = 1, ..., N; \emptyset \to \emptyset, 1 \to 1].$$

The meaning of this set of rules is that the state letter P_j can freely move (if the $(\bar{\mathcal{E}}, \Omega)$-coordinates are $(\emptyset, 1)$, and L_j-letters and R_j-letters stay next to K_j-letters). Indeed, if in an admissible word, the $\mathcal{A}$-letter next to the right of P_j is a_i (and rule $\tau(1, \emptyset, i)$ is applicable to the word) then applying the rule $\tau(1, \emptyset, i)$, we move all P_j one letter to the right. Similarly, if we want to move P_j one letter to the left, we need to apply $\tau(1, \emptyset, i)^{-1}$ for an appropriate i.

Notice that, for every $z \in \mathcal{K}$, all state letters z_j, $j = 1, ..., N$ behave in the same way when we apply $\tau(1, \emptyset, i)$. This important property will hold for all other rules of S.

Recall also that the two a_i in $a_i P_j a_i^{-1}$ are not the same: they are brothers taken from different alphabets (see Section 2.4).

The set $S^+(12)$ consists of one rule $\tau(12, r)$ for each $r \in \bar{\mathcal{E}} \setminus \{\emptyset\}$:

$$[\overleftarrow{L}_j \overset{\ell}{\to} \overleftarrow{L}_j, R_j \overset{\ell}{\to} R_j, j = 1, ..., N; \emptyset \to r, 1 \to 2]$$

This rule does not insert any tape letters, it simply changes the $\bar{\mathcal{E}}$- and Ω-coordinates of state letters. This *transition rule* prepares the machine for step 2.

The set $S^+(2)$ consists of rules $\tau(2, r, i)$ where $r \in \bar{\mathcal{E}}$, $i = 1, ..., \bar{m}$:

$$\tau(2, r, i) = [L_j \to a_i L_j a_i^{-1}, R_j \overset{\ell}{\to} R_j, j = 1, ..., N; r \to r, 2 \to 2].$$

The meaning of these rules is that they allow the state letters L_j to move freely while R_j stays next to $\vec{R}_j$.

The set $\mathcal{S}^+(23)$ consists of one rule for each $r \in \bar{\mathcal{E}}$:

$$\tau(23, r) = [L_j \xrightarrow{\ell} L_j, R_j \xrightarrow{\ell} R_j; r \to r, 2 \to 3].$$

The meaning of this rule is that the machine can start step 3 when L_j and P_j meet (that is when the inner parts of $L_j z$-sectors are empty).

The set $\mathcal{S}^+(3)$ consists of one rule for each $r \in \bar{\mathcal{E}}$ and each i from 1 to $\bar{m}$:

$$\tau(3, r, i) = [L_j \xrightarrow{\ell} L_j, R_j \to a_i^{-1} R_j a_i, j = 1, ..., N; r \to r, 3 \to 3].$$

These rules allow the state letter R_j to move freely between P_j and $\vec{R}_j$.

The set $\mathcal{S}^+(34)$ consists of one rule for each non-empty $r \in \bar{\mathcal{E}}$:

$$\tau(34, r) = [L_j \xrightarrow{\ell} rL_j, P_j \xrightarrow{\ell} P_j, j = 1, ..., N; r \to r, 3 \to 4].$$

This rule can be applied when the state letters L_j, P_j, R_j meet together; it inserts r to the left of the state letters L_j, and prepares the machine for step 4.

The set $\mathcal{S}^+(4)$ consists of rules $\tau(4, r, i)$, $r \in \bar{\mathcal{E}}, i = 1, ..., \bar{m}$:

$$\tau(4, r, i) = [L_j \to a_i L_j a_i^{-1}, P_j \xrightarrow{\ell} P_j, j = 1, ..., N; r \to r, 4 \to 4].$$

These rules allow the state letter L_j to move freely between $\overleftarrow{L}_j$ and P_j.

The set $\mathcal{S}^+(45)$ consists of one rule for each $r \in \bar{\mathcal{E}}$:

$$\tau(45, r) = [\overleftarrow{L}_j \xrightarrow{\ell} \overleftarrow{L}_j, P_j \xrightarrow{\ell} P_j, j = 1, ..., N; r \to r, 4 \to 5].$$

The machine can start step 5 when L_j and $\overleftarrow{L}_j$ meet.

The set $\mathcal{S}^+(5)$ consists of one rule for each $r \in \bar{\mathcal{E}}, i = 1, ..., \bar{m}$:

$$\tau(5, r, i) = [\overleftarrow{L}_j \xrightarrow{\ell} \overleftarrow{L}_j, R_j \to a_i^{-1} R_j a_i, j = 1, ..., N; r \to r, 5 \to 5].$$

These rules allow R_j move freely between P_j and $\vec{R}_j$.

Finally the set $\mathcal{S}^+(51)$ consists of one rule for each $r \in \bar{\mathcal{E}}$:

$$\tau(51, r) = [\overleftarrow{L}_j \xrightarrow{\ell} \overleftarrow{L}_j, R_j \xrightarrow{\ell} R_j, j = 1, ..., N; r \to \emptyset, 5 \to 1].$$

The cycle is complete, the machine can start step 1 again when $\overleftarrow{L}_j$ meets L_j, and R_j meets $\vec{R}_j$.

Rules from the set $\mathcal{S}(12) \cup \mathcal{S}(23) \cup \mathcal{S}(34) \cup \mathcal{S}(45) \cup \mathcal{S}(51)$ will be called *transition rules*.

A graphical description of the work of the S-machine $\mathcal{S}$ is presented on Figure 5 below.

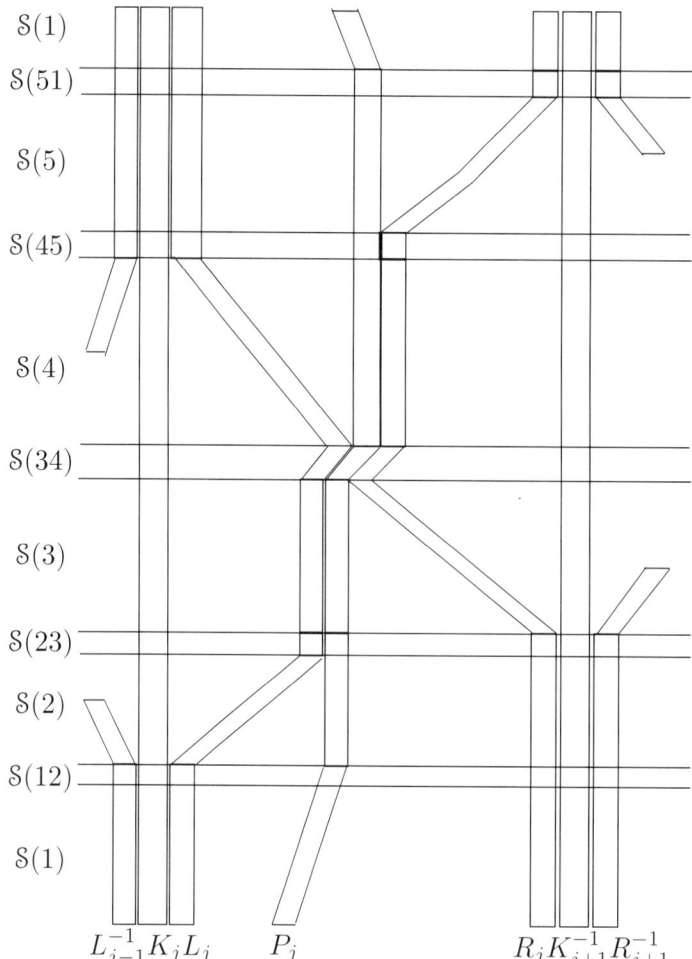

Figure 5.

Some of the main properties of $\mathcal{S}$ are the following.

Remark 2.1. *Notice that if $W = y_1 u_1 y_2 ... y_{t+1}$ is an admissible word for $\mathcal{S}$, y_1 (resp. y_{t+1}) is $K_j(r,i)^{\pm 1}$ for some j, r, i then one does not need trimming the prefix (resp. suffix) in order to produce $W \circ \tau$ because in this case $v((y_1)_-)$ (resp. $u(y_{t+1})$) is empty for all rules τ.*

For every word w in $\{a_1, ..., a_{\bar{m}}\}$ let us denote the word

$$K_1(\emptyset, 1) L_1(\emptyset, 1) P_1(\emptyset, 1) w(P_1) R_1(\emptyset, 1) K_2(\emptyset, 1)^{-1} ...$$
$$K_N(\emptyset, 1)^{-1} R_N(\emptyset, 1)^{-1} w(P_N)^{-1} P_N(\emptyset, 1)^{-1} L_N(\emptyset, 1)^{-1} \quad (2.3)$$

by $\Sigma(w)$. Notice that $\Sigma(\emptyset) = \Sigma$. Clearly, $\Sigma(w) K_1(\emptyset, 1)$ is admissible for $\mathcal{S}$ for every positive w.

Lemma 2.2. *An admissible word for $\mathcal{S}$ cannot have $L_j^{-1} L_j$-, $P_j^{-1} P_j$-, or $R_j R_j^{-1}$-sectors.*

Proof. Indeed if W is an admissible word, then W^{-1} is also an admissible word. So the inner part of a $L_j^{-1}L_j$-, $P_j^{-1}P_j$-, or $R_jR_j^{-1}$-sector of W must be both positive and negative. Hence the inner part must be empty, and W is not reduced, a contradiction. $\square$

Lemma 2.3. *Suppose that a positive word w' in $\{a_1, ..., a_{\bar{m}}\}$ can be obtained from a positive word w by insertion or deletion of a word $r \in \bar{\mathcal{E}}$. Then there exists a sequence h of rules of $\mathcal{S}$ such that*

$$\Sigma(w)K_1(\emptyset, 1) \circ h = \Sigma(w')K_1(\emptyset, 1). \tag{2.4}$$

The length of h is bounded by a linear function in $|w| + |w'|$.

Thus, in particular, for every positive word w which is equal to 1 in $\bar{\mathcal{G}}$ there exists a word h in $\mathcal{S}$ such that

$$\Sigma K_1(\emptyset, 1) \circ h = \Sigma(w)K_1(\emptyset, 1).$$

Proof. Without loss of generality we can assume that w' is obtained from w by inserting r. Then we can first apply to $\Sigma(w)K_1(\emptyset, 1)$ rules from $\mathcal{S}(1)$ to move the letters $P_j^{\pm 1}$ to the places in copies of $w^{\pm 1}$ where the insertion will occur ($j = 1, ..., N$). Then a rule from $\mathcal{S}(12)$ (which is applicable) will change the coordinates from $(\emptyset, 1)$ to $(r, 2)$. After that we can apply rules from $\mathcal{S}(2)$ to move $L_j^{\pm 1}$ next to $P_j^{\pm 1}$ and ($j = 1, ..., N$). Then a rule from $\mathcal{S}(23)$ will change the coordinates to $(r, 3)$. After that we can apply rules from $\mathcal{S}(3)$ to move $R_j^{\pm 1}$ next to $P_j^{\pm 1}$. Then a rule from $\mathcal{S}(34)$ will insert r (resp. r^{-1}) to the left of L_j (resp. to the right of L_j^{-1}). Then rules from $\mathcal{S}(4)$ will move L_j back to K_j, for odd j and L_j^{-1} to K_{j+1} for even j. Then a rule from $\mathcal{S}(45)$ will change the coordinates to $(r, 5)$, and rules from $\mathcal{S}(5)$ will move the letters R_j back to K_{j+1}^{-1} (j odd), and R_j^{-1} to K_j^{-1} (j even). Finally we can apply a rule from $\mathcal{S}(51)$ and obtain $\Sigma(w')K_1(\emptyset, 1)$ as required.

It is clear that the number of rules applied in the described process is bounded by a linear function in $|w| + |w'|$.

The "in particular" statement immediately follows from the first statement and Lemma 2.1. $\square$

2.6 The S-machines $\bar{\mathcal{S}}$ and $\mathcal{S} \cup \bar{\mathcal{S}}$

The S-machine $\bar{\mathcal{S}}$ is similar to $\mathcal{S}$ so we only present the differences between $\mathcal{S}$ and $\bar{\mathcal{S}}$.

We introduce disjoint copies of sets $\mathcal{K}$, $\tilde{\mathcal{K}}$, $\mathcal{A}$, denote them by $\bar{\mathcal{K}}$, $\bar{\tilde{\mathcal{K}}}$, $\bar{\mathcal{A}}$, respectively. In order to make $\mathcal{S}$ and $\bar{\mathcal{S}}$ "communicate", we identify $z_j(\emptyset, 1)$ with $\bar{z}_j(\emptyset, 1)$, $z \in \{K, L, P, R\}$, $j = 1, ..., N$, and $a_i(P_j)$ with $\bar{a}_i(\bar{P}_j)$ $i = 1, ..., m, j = 1, ..., N$. Notice that we do not identify $a_{m+1}, ..., a_{\bar{m}}$ with their "bar"-brothers $\bar{a}_{m+1}, ..., \bar{a}_{\bar{m}}$, and we do not identify $a_i(z)$ with $\bar{a}_i(z)$ for $z \neq P_j$. Notice also that this identification of letters makes the word $\bar{\Sigma}$ coincide with the hub Σ.

Admissible words of $\bar{S}$ have the same form as admissible words of S except for the following differences:

- All letters are replaced by their "bar"-brothers.

- We drop the restriction that certain parts of admissible words are positive (negative).

- The $\bar{K}_1 z$-, $\bar{L}_1 z$-, $\bar{P}_1 z$- and $\bar{R}_1 z$-sectors must be empty ($z \in \bar{\mathcal{K}}$). Thus in every admissible word of $\bar{S}$ the part between $\bar{K}_1$ and $\bar{K}_2^{-1}$ contains no $\bar{\mathcal{A}}$-letters.

The rules of $\bar{S}$ are obtained from the corresponding rules of S by replacing every letter by its "bar"-brother and by removing letters from $\bar{\mathcal{A}}(\bar{z}_1)$, $z \in \{K, L, P, R\}$. Thus rules of $\bar{S}$ do not insert any $\bar{\mathcal{A}}$-letters in the interval between $\bar{K}_1$ and $\bar{K}_2^{-1}$, and work in other parts of the admissible words just as S.

For every admissible word W for S, we define an admissible word $\bar{W}$ of $\bar{S}$ by adding *bar* to all letters, and removing non-$\mathcal{K}$-letters from all zz'-sectors, such that z or $(z')^{-1}$ belongs to $\{K_1, L_1, P_1, R_1, K_2\}$.

Lemma 2.4. *Suppose that w, u are reduced words in $\{a_1, ..., a_{\bar{m}}\}$, and w' is the freely reduced form of $wuru^{-1}$, $r \in \bar{\mathcal{E}}$. Then there exists a sequence h of rules of $\bar{S}$ such that*

$$\bar{\Sigma}(w) K_1(\emptyset, 1) \circ h = \bar{\Sigma}(w') K_1(\emptyset, 1). \tag{2.5}$$

Therefore, for every word w which is equal to 1 in $\bar{\mathcal{G}}$ there exists a word h in S such that

$$\Sigma K_1(\emptyset, 1) \circ h = \bar{\Sigma}(w) K_1(\emptyset, 1).$$

If all letters in w are with indices $\leq m$, then one can remove $^-$ from the right-hand side of the previous formula (because we have identified these letters with their "bar"-brothers).

Proof. The proof is similar to the proof of Lemma 2.3. The only difference is that at the beginning we transform every subword of the form $P_j(\emptyset, 1) w(P_j) R_j(\emptyset, 1)$ into $w(L_j) q(L_j) P_j q(L_j)^{-1} R_j(\emptyset, 1)$ using rules from $\bar{S}(1)$ instead of $S(1)$. □

We shall also need the combined S-machine $S \cup \bar{S}$. An admissible word for $S \cup \bar{S}$ is either an admissible word for $\bar{S}$ or a word satisfying all conditions from the definition of an admissible word for S except for the positivity condition. The set of rules of $S \cup \bar{S}$ is the union of the set of rules of S and $\bar{S}$.

For the sake of simplicity of notation, for every word W we denote the projection $W_{\mathcal{A} \cup \bar{\mathcal{A}}}$ of W onto the alphabet $\mathcal{A} \cup \bar{\mathcal{A}}$ by W_a.

THE CONJUGACY PROBLEM AND HIGMAN EMBEDDINGS

We shall call a rule $\tau \in \mathcal{S} \cup \bar{\mathcal{S}}$ *active with respect to* zz'-*sectors* if $u_\tau(z) \neq \emptyset$ or $v_\tau(z'_-) \neq \emptyset$.

We define a map β of the subset of generators $\mathcal{A} \cup \bar{\mathcal{A}}$ of the group $\mathcal{H}$ onto $\{a_1, \ldots, a_{\bar{m}}\} \cup \{1\}$ by the rule $\beta(a_i(z)) = \beta(\bar{a}_i(z)) = a_i$ for all the $\mathcal{A}$- and $\bar{\mathcal{A}}$-letters. Hence the β-image of an arbitrary word in $\mathcal{A} \cup \bar{\mathcal{A}}$ is defined as an element of the free group with basis $\{a_1, \ldots, a_{\bar{m}}\}$.

The following two properties of $\mathcal{S} \cup \bar{\mathcal{S}}$ immediately follows from its definition.

Lemma 2.5. *If a rule $\sigma \in \mathcal{S}(\omega)$ (resp. $\sigma \in \bar{\mathcal{S}}(\omega)$) is active with respect to zz'-sectors then every rule in $\mathcal{S}(\omega)$ (resp., $\bar{\mathcal{S}}(\omega)$) is active with respect to zz'-sectors.*

Lemma 2.6. *Let h be a sequence of rules from $\mathcal{S} \cup \bar{\mathcal{S}}$ applicable to an admissible (for $\mathcal{S} \cup \bar{\mathcal{S}}$) word W. Then for every zz'-sector $z(r,i)W'z'(r,i)$ of W there exist two words $u = u(h)$ and $v = v(h)$ such that $z(r,i)W'z'(r,i) \circ h = z(r_1, i_1)uW'vz'(r_1, i_1)$ in the free group for some coordinates r_1, i_1. The words $u = u(z,h)$ and $v = v(z',h)$ depend on z, z' and h only (i.e. they do not depend on the inner part of W'). Moreover we have the following equalities:*

$$u(K_j, h) \equiv v(K_j, h) \equiv \emptyset, \beta(u(P_j, h)) = \beta(v(P_j, h))^{-1},$$
$$\beta(u(R_j, h)) \equiv \beta(v(R_j, h))^{-1}.$$

We also have that $\beta(u(L_j, h)) = \beta(v(L_j, h))^{-1}$ provided h contains no rules from $\mathcal{S}(34)$.

Lemma 2.7. *Let U be any word in $\tilde{\mathcal{K}}$, $i \in \{1,2,3,4,5\}$. Then there exist two sequences of numbers $\{\varepsilon_z, \delta_z \in \{-1, 0, 1\} \mid z \in \tilde{\mathcal{K}}\}$, such that for every admissible word $W = W(r,i)$ with base U, and any group word w in the alphabet $\{a_1, \ldots, a_{\bar{m}}\}$ there exists a word h in $\bar{\mathcal{S}}(i)$ satisfying the following properties:*

1. *h is a copy[3] of w in the alphabet $\bar{\mathcal{S}}(i)$;*

2. *W is in the domain of h;*

3. *$W \circ h$ is obtained from W by all replacements of the form*

$$\bar{z}(r,i) \to w(z_-)^{\varepsilon_z} \bar{z}(r,i) w(z)^{\delta_z},$$

where z runs over all letters in the base of W. Here $w(z)$ and $w(z_-)$ are copies of w in the alphabets $\bar{\mathcal{A}}(z)$ and $\bar{\mathcal{A}}(z_-)$.

Now we need to explain how to simulate the S-machines $\mathcal{S}$ and $\bar{\mathcal{S}}$ in a group $\mathcal{H}$.

---

[3]Here and below, we say that a word v is a *copy* of a word u if v is obtained from u by replacing letters by their images under a bijective map of the alphabet.

2.7 Converting S into a set of relations

In order to convert the machine S into a list of defining relations, we need to introduce one more set of letters $\mathcal{X} = \{x(a,\tau) \mid a \in \mathcal{A} \backslash \cup_j \mathcal{A}(P_j), \tau \in S^+\}$. We will subdivide this set into subsets $\mathcal{X}(z,\tau) = \{x(a,\tau) \mid a \in \mathcal{A}(z)\}$, $z \in \tilde{\mathcal{K}}, z \ne P_j (j = 1, ..., N), \tau \in S$ and into subsets $\mathcal{X}(z) = \cup_{\tau \in S} \mathcal{X}(z,\tau)$, $z \in \tilde{\mathcal{K}}$, and $\mathcal{X}(\tau) = \cup_{z \in \tilde{\mathcal{K}}} \mathcal{X}(z,\tau)$, $\tau \in S^+$. We also set $\mathcal{X}(z^{-1}) = \mathcal{X}(z_-)$ for every $z \in \tilde{\mathcal{K}}$.

For every $\tau \in S$, we also consider map α_τ from $\mathcal{A}$ to the free group generated by $\mathcal{A}$ and $\mathcal{X} \cup \mathcal{X}^{-1}$. Let $\tau \in S^+$, $a \in \mathcal{A}(z)$, $z \in \tilde{\mathcal{K}}$. Then:

$$\alpha_\tau(a) = \begin{cases} x(a,\tau)a & \text{if } z = K_j, j = 1, ..., N, \\ x(a,\tau)a & \text{if } z = L_j, j = 1, ..., N, \\ a & \text{if } z = P_j, j = 1, ..., N, \\ ax(a,\tau) & \text{if } z = R_j, j = 1, ..., N. \end{cases}$$

The definition of α_τ^{-1}, $\tau \in S^+$, is obtained from the definition of α_τ by replacing all $\mathcal{X}$-letters by their inverses.

We expand these maps to all words in the alphabet $\mathcal{A} \cup \mathcal{A}^{-1} \cup \mathcal{K} \cup \mathcal{K}^{-1}$ in the natural way (mapping letters from $\mathcal{K} \cup \mathcal{K}^{-1}$ to themselves). Let U be any word in $\mathcal{A}$. Then U is obviously equal to the *projection* of $\alpha_\tau(U)$ onto the alphabet $\mathcal{A}$, i.e. the word obtained from $\alpha_\tau(U)$ by removing all non-$\mathcal{A}$-letters. In general, the projection of a word U onto an alphabet Y will be denoted by U_Y.

Also with every $\tau \in S^+$, we associate a set of letters

$$\Theta(\tau) = \{\theta(\tau, z) \mid z \in \{K_j, L_j, P_j, R_j\}, j = 1, ..., N\}.$$

Let $\Theta(z) = \{\theta(\tau, z) \mid \tau \in S^+\}$, $z \in \{K_j, L_j, P_j, R_j\}$. Finally the union of all $\Theta(\tau)$ is denoted by Θ.

Let $\tau = [U_1 \to V_1, ..., U_k \to V_k; r \to r', l \to l']$ be one of the rules of S^+. Let $u(z), v(z)$, $z \in \tilde{\mathcal{K}}$, be the words associated with this rule as in Section 2.4.

Then we associate with τ the following list of *main relations* (or just (Θ, k)-*relations*):

$$\theta(\tau, z_-)^{-1} z(r, l) \theta(\tau, z) = \alpha_{\tau^{-1}}(v_i) z(r', l') \alpha_{\tau^{-1}}(u_i), i = 1, ..., n, \qquad (2.6)$$

(Thus, the replacements associated with the rule τ are simulated by "conjugation". Figure 6 shows a relation involving a P_j-letter, corresponding to a rule $\tau = \tau(1, \emptyset, i)$; there $P = P_j(1,0)$, $a = a_i(P_j)$, $a' = a_i(L_j)$, $x' = x(a_i(L_j), \tau)$, $\theta = \theta(\tau, P_j)$, $\theta' = \theta(\tau, L_j)$.)

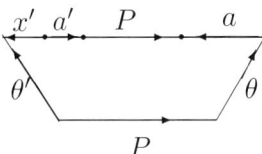

Figure 6.

We also need the following *auxiliary relations*. Let $z \in \{K_j, L_j, P_j, R_j\}$, $j = 1, ..., N$, $a, b \in \mathcal{A}(z)$. Then we add *auxiliary Θ-relations* (or (θ, a)-*relations*):

$$\theta(\tau, z)^{-1} \alpha_\tau^+(a) \theta(\tau, z) = \alpha_\tau^-(a) \text{ if } \tau \text{ does not lock } zz_+\text{-sector} \qquad (2.7)$$

(On Figure 7, $a = a_i(L_j)$, $x = x(a, \tau)$, $\theta = \theta(\tau, L_j)$ where $\tau = \tau(2, r, i')$ for some i, i', r.)

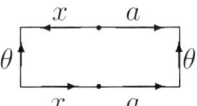

Figure 7.

and *auxiliary (a, x)-relations*:

$$\begin{aligned} ax(b, \tau) a^{-1} &= x(b, \tau)^4 &&\text{if } z = K_j \text{ or } z = L_j, \\ a^{-1} x(b, \tau) a &= x(b, \tau)^4 &&\text{if } z = R_j. \end{aligned} \qquad (2.8)$$

(On Figure 8, $a = a_i(L_j)$, $x = x(a_{i'}(L_j), \tau)$ where $\tau = \tau(2, r, i')$ for some i, j, i', r.)

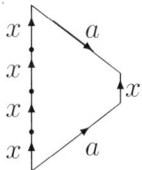

Figure 8.

Finally let b' be the "brother" of b in the alphabet $\mathcal{A}(z_-)$ if $z = K_j$ or $z = L_j$. Then we need auxiliary (k, x)-*relations*:

$$\begin{aligned} z(r, i) x(b, \tau) &= x(b', \tau) z(r, i) &&\text{if } z = K_j, \\ z(r, i) x(b, \tau) &= x(b', \tau)^4 z(r, i) &&\text{if } z = L_j. \end{aligned} \qquad (2.9)$$

(On Figure 9, $L = L_j(0, 1)$, $x = x(a_i(L_j), \tau)$, $x' = x(a_i(\overleftarrow{L}_j), \tau)$, where $\tau = \tau(2, r, i')$ for some i, j, r, i'.)

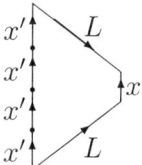

Figure 9.

Let $\mathcal{R}(\mathcal{S})$ be the set of all relations (2.6) - (2.9) corresponding to the set of S-rules $\mathcal{S}^+$. It is easy to see that the relations corresponding to the negative rules from $\mathcal{S}$ follow from $\mathcal{R}(\mathcal{S})$.

2.8 Converting $\bar{\mathcal{S}}$ into a set of relations

To convert $\bar{\mathcal{S}}$ into a set of relations we do not need $\mathcal{X}$-letters, but we need "bar"-brothers of the letters from Θ. Let this set be denoted by $\bar{\Theta}$. The easiest way to explain how to convert an S-rule of $\bar{\mathcal{S}}$ to a set of relations is the following: convert it as in Section 2.7, then add bars $^-$ to all letters, replace all letters from $\mathcal{X}$ and from $\cup \bar{A}(z_1), z \in \{K, L, P, R\}$) by 1. Let $\mathcal{R}(\bar{\mathcal{S}})$ be the set of all relations corresponding to $\bar{\mathcal{S}}^+$.

2.9 The presentation of the group $\mathcal{H}$

The group $\mathcal{H}$ which is the main object studied in this paper is given by the set of generators $\mathcal{K} \cup \bar{\mathcal{K}} \cup \mathcal{A} \cup \bar{\mathcal{A}} \cup \Theta \cup \bar{\Theta} \cup \mathcal{X}$ and the set of relations $\mathcal{R} = \mathcal{R}(\mathcal{S}) \cup \mathcal{R}(\bar{\mathcal{S}}) \cup \{\Sigma\}$.

3 The first properties of $\mathcal{H}$

3.1 Diagrams

Recall the well-known van Kampen-Lyndon topological interpretation of the consequences of defining relations in groups [LS], [Ol1]. We define a van Kampen *diagram* over some group presentation $\mathcal{L} = \langle x_1, \ldots, x_k \mid \mathcal{P} \rangle$ (or briefly, by misuse of language, over the group $\mathcal{L}$), where the defining words from $\mathcal{P}$ are cyclically reduced. It is an oriented, connected, and simply connected, finite planar 2-complex such that every oriented edge e of it has a *label* $\phi(e) \equiv x_i^{\pm 1}$, $i = 1, \ldots, k$, where $\phi(e^{-1}) \equiv \phi(e)^{-1}$. In addition, the word read on the boundary $\partial \Pi$ of an arbitrary 2-cell Π (we call it a *cell* latter on), coincides with one of defining word, up to a cyclic permutation and an inversion.

According to the van Kampen lemma a word $w = w(x_1, \ldots, x_k)$ equals 1 in a group $\mathcal{L}$ if and only if there exists a van Kampen diagram Δ over

$\mathcal{L}$ with w read clockwise on the boundary $\partial(\Delta)$ starting from some vertex. Full details can be found in [LS] or [Ol1].

An *annular* diagram Δ is not simply connected (unlike a van Kampen diagram), but its complement on the plane has two components: the inner and the outer. So Δ has the *inner* and *outer* contours. By the Schupp lemma [LS], [Ol1] their labels, read clockwise, are conjugate in $\mathcal{L}$. Conversely, for any pair of words u, v, which represent non-trivial conjugate elements of $\mathcal{L}$, there exists an annular diagram whose contours are labelled by u and v respectively.

3.2 $\mathcal{G}$ embeds into $\mathcal{H}$

We define a map δ of the set of generators of the group $\mathcal{H}$ into the group $\bar{\mathcal{G}}$ by setting $\delta(a_i(z)) = \delta(\bar{a}_i(z)) = a_i$ for all the $\mathcal{A}$- and $\bar{\mathcal{A}}$-letters and mapping other letters from the whole set of generators to 1.

Lemma 3.1. *The map δ extends to the epimorphism $\delta : \mathcal{H} \to \bar{\mathcal{G}}$.*

Proof. The map δ can be considered as an endomorphism of the free group over the generators of $\mathcal{H}$. It immediately follows from the list of relations of the group $\mathcal{H}$, that the δ-image of every defining word for $\mathcal{H}$ vanishes in $\bar{\mathcal{G}}$. Indeed, it is obvious for the hub word Σ since it does not contain $\mathcal{A} \cup \bar{\mathcal{A}}$-letters. It is also true for all auxiliary relations (2.7), (2.8), (2.9) because each of them contains exactly two occurrences of $\mathcal{A} \cup \bar{\mathcal{A}}$-letters whose δ-images are mutually inverse. Same is true for all relations (2.6) corresponding to S-rules τ and $z \in \tilde{\mathcal{K}}$ unless $\tau = \tau(34, r)$ and $z = L_j$ or $\tau = \bar{\tau}(34, r)$ and $z = \bar{L}_j$ (for some r and j). In the latter case the δ-image of the relator (2.6) is equal to r which belongs to $\bar{\mathcal{E}}$ and is equal to 1 in $\bar{\mathcal{G}}$. □

The following lemma is similar to Lemma 3.1 and immediately follows from the definition of $\mathcal{S} \cup \bar{\mathcal{S}}$.

Lemma 3.2. *For every word U in the domain of $\tau \in \mathcal{S} \cup \bar{\mathcal{S}}$, $\delta(U) = \delta(U \circ \tau)$ in $\bar{\mathcal{G}}$.*

Lemma 3.3. *a) Let W be an admissible word for the S-machine $\mathcal{S}$, z be its first $\mathcal{K}$-letter and z' be its last $\mathcal{K}$-letter. Let $\tau \in \mathcal{S}$, and let $v = v(z_-)$ and $u = u(z')$ be the words associated with τ, defined in Section 2.4. Suppose that $\tau \in \mathcal{S}$ is applicable to W. Then*

$$\theta(\tau, z_-)^{-1}\alpha_\tau(W)\theta(\tau, z') = \alpha_{\tau^{-1}}(v)\alpha_{\tau^{-1}}(W \circ \tau)\alpha_{\tau^{-1}}(u)$$

in $\mathcal{H}$.

b) Let W be an admissible word for the S-machine $\bar{\mathcal{S}}$, z is its first $\mathcal{K}$-letter and z' is its last $\mathcal{K}$-letter. Let $\tau \in \bar{\mathcal{S}}$, and $v = v(z_-)$ and $u = u(z')$ be the words associated with τ, defined in Section 2.4. Suppose that τ is applicable to W. Then

$$\theta(\tau, z_-)^{-1}W\theta(\tau, z') = v \cdot (W \circ \tau) \cdot u$$

in $\mathcal{H}$.

Proof. The proof follows from the definition of admissible words for $\mathcal{S}$ and $\bar{\mathcal{S}}$ (we do not need the positivity of certain segments of admissible words of $\mathcal{S}$), relations (2.6), (2.7), and similar relations for $\bar{\mathcal{S}}$. □

Definition 3.1. Let w be a word over $\{a_1, ..., a_{\bar{m}}\}$. For every $z \in \tilde{\mathcal{K}}$, let $w(z)$ be the word w rewritten in the alphabet $\mathcal{A}(z)$. Let w_1, w_2, w_3, w_4 be words over $\{a_1, ..., a_{\bar{m}}\}$, let $j = 1, ..., N$. Denote the word

$$w_1(\overleftarrow{L}_j) L_j w_2(L_j) P_j w_3(P_j) R_j w_4(R_j)$$

by $W_j = W_j(w_1, w_2, w_3, w_4)$. Denote the word

$$K_1 W_1 K_2^{-1} W_2^{-1} ... K_N^{-1} W_N^{-1}$$

by $\Sigma(w_1, w_2, w_3, w_4)$. Let the word $\bar{\Sigma}(w_1, w_2, w_3, w_4)$ be obtained from the word $\Sigma(w_1, w_2, w_3, w_4)$ by replacing W_1 with $L_1 P_1 R_1$, and adding ¯ to all letters. Finally for every $i \in \{1, 2, 3, 4, 5\}$ and $r \in \bar{\mathcal{E}}$ let us denote the word $\Sigma(w_1, w_2, w_3, w_4)(r, i)$ by $\Sigma_{r,i}(w_1, w_2, w_3, w_4)$ and $\bar{\Sigma}(w_1, w_2, w_3, w_4)(r, i)$ by $\bar{\Sigma}_{r,i}(w_1, w_2, w_3, w_4)$.

Notice that if w_1, w_2, w_4 are positive words (we do not care about w_3) then the word $\Sigma_{r,i}(w_1, w_2, w_3, w_4) K_1(r, i)$ (see the definition of $U(r, i)$ in 2.3) is admissible for $\mathcal{S}$ for every r, i. The word $\bar{\Sigma}_{r,i}(w_1, w_2, w_3, w_4) \bar{K}(r, i)$ is admissible for $\bar{\mathcal{S}}$ for every w_1, w_2, w_3, w_4, r, i.

Lemma 3.4. *Preserving the notation from Definition 3.1, suppose that w_1, w_2, w_4 are positive words. Let $j \in \{1, ..., N\}$, $r \in \bar{\mathcal{E}}$, $i \in \{1, 2, 3, 4, 5\}$, $W = W_j(r, i)$, and τ applicable to W. Then there exist words X_1, X_1', X_2, and X_2' in the alphabet $\mathcal{X}(\tau)$, depending on w_1, w_2, w_4 (but not on w_3) such that*

a) $X_1 W X_1' = \alpha_\tau(W)$;

b) $X_2 W X_2' = \alpha_{\tau^{-1}}(W)$;

c) the words X_1 and X_2 contain only letters from the $\mathcal{X}(\overleftarrow{L}_j, \tau)$, the words X_1', X_2' contain letters only from $\mathcal{X}(R_j, \tau)$.

The lengths of X_i, X_i' ($i = 1, 2$) are bounded by a recursive function in $|w_1| + |w_2| + |w_4|$ (in fact it does not exceed $4^{|w_1|+|w_2|+|w_4|+1}$).

Proof. The statement immediately follows from relations (2.8). □

Lemma 3.5. *a) In the above notation, suppose that w_1, w_2, w_4 are positive words, $r \in \bar{\mathcal{E}}, i \in \{1, 2, 3, 4, 5\}$. Further suppose that $\tau \in \mathcal{S}$ is applicable to*

$$\Sigma_{r,i}(w_1, w_2, w_3, w_4) K_1(r, i),$$

THE CONJUGACY PROBLEM AND HIGMAN EMBEDDINGS 25

and τ changes the coordinates (r,i) into (r',i'). Then, by (2.6) - (2.9),

$$X^{-1}\Sigma_{r,i}(w_1,w_2,w_3,w_4)XK_1(r',i') = (\Sigma_{r,i}(w_1,w_2,w_3,w_4)K_1(r,i))\circ\tau$$

for some word $X = X_1\theta X_2$ where X_1 and X_2 are words in $\mathfrak{X}(\tau)$, whose lengths are bounded by a recursive function in $|w_1|+|w_2|+|w_4|$, and $\theta \in \Theta(\tau,(K_1)_-)$.

b) Suppose that w_1,w_2,w_3,w_4 are arbitrary words, and τ is applicable to

$$\bar\Sigma_{r,i}(w_1,w_2,w_3,w_4)\bar K_1(r,i).$$

Then for some $\theta \in \bar\Theta(\tau,(K_1)_-)$ we have

$$\theta^{-1}\bar\Sigma_{r,i}(w_1,w_2,w_3,w_4)\theta K_1(r',i') = (\bar\Sigma_{r,i}(w_1,w_2,w_3,w_4)K_1(r,i))\circ\tau.$$

Proof. a) Fix a number j between 1 and N. Let $K = \overleftarrow{L}_j(r,i)$, $K' = \overrightarrow{R}_j(r,i)$. Since τ is applicable to $\Sigma_{r,i}(w_1,w_2,w_3,w_4)K_1(r,i)$, it is applicable to $KW_j(w_1,w_2,w_3,w_4)K'$. Hence by Lemma 3.4 a), there exist words X_j, X'_j in the alphabets $\mathfrak{X}(K_j,\tau)$ and $\mathfrak{X}(R_j,\tau)$ respectively such that

$$X_j KW_j K' X'_j = K\alpha_\tau(W_j)K'. \tag{3.10}$$

The words X_j and X'_j depend only on w_1,w_2,w_4 and τ. So all X_j (resp. X'_j), $j=1,...,N$, are copies of each other. The word X_j contains letters from the set $\mathfrak{X}(K_-,\tau)$ only, the word X'_j contains letters from $\mathfrak{X}(K',\tau)$ only. Hence by relations (2.9) we have

$$KX_j = X_{j-1}K, \quad (X'_j)^{-1}K' = K'(X'_{j+1})^{-1}. \tag{3.11}$$

Using equalities (3.10) and (3.11) together with the definition of the word $\Sigma_{r,i}(w_1,w_2,w_3,w_4)$, we get

$$X_N^{-1}\Sigma_{r,i}(w_1,w_2,w_3,w_4)X_N K_1(r,i) = \alpha_\tau(\Sigma_{r,i}(w_1,w_2,w_3,w_4))K_1(r,i). \tag{3.12}$$

Now using Lemma 3.3, relation (2.6), and the fact that $u(K_1) = v(K_1) = \emptyset$ (Remark 2.1) we deduce that

$$\theta^{-1}\alpha_\tau(\Sigma_{r,i}(w_1,w_2,w_3,w_4))\theta K_1(r',i') = \alpha_{\tau^{-1}}(\Sigma_{r,i}(w_1,w_2,w_3,w_4)K_1(r,i)\circ\tau) \tag{3.13}$$

where $\theta = \theta(\tau,(K_1)_-)$.

Now using Lemma 3.4 b), c), we deduce (similarly to (3.12)) that for some word Y_N in $\mathfrak{X}((K_1)_-,\tau)$, we have

$$Y_N^{-1}\alpha_{\tau^{-1}}(\Sigma_{r',i'}(w_1,w_2,w_3,w_4)\circ\tau)Y_N K_1(r',i') = (\Sigma_{r,i}(w_1,w_2,w_3,w_4)K_1(r,i))\circ\tau. \tag{3.14}$$

Finally combining (3.12), (3.13) and (3.14), we deduce statement a) of the lemma with $X = X_N\theta Y_N$.

Statement b) immediately follows from Lemma 3.3 b). $\square$

Let $\mathcal{H}_0$ be the group given by all defining relations of $\mathcal{H}$ except for the hub Σ.

Lemma 3.6. *Assume a positive word $w' = w'(a_1, \ldots, a_{\bar{m}})$ can be obtained from a positive word w by an insertion (deletion) of a word $r \in \bar{\mathcal{E}}$. Then the words $\Sigma(w)$ and $\Sigma(w')$ are conjugate in the group $\mathcal{H}_0$. There is an annular diagram for this conjugacy without $\bar{\theta}$-cells, in which the contours p and p' can be connected by a path whose length is bounded by a recursive function of $|w| + |w'|$.*

Proof. Indeed, by Lemma 2.3 there exists a word h in the alphabet $\mathcal{S}$ which is applicable to $\Sigma(w)K_1(\emptyset, 1)$ and such that

$$(\Sigma K_1(\emptyset, 1)) \circ h = \Sigma(w')K_1(\emptyset, 1).$$

The length of h is bounded by a recursive function of $|w| + |w'|$.

Notice that for every words w_1, w_2, w_3, w_4 where w_1, w_2, w_4 are positive and for every rule τ applicable to $\Sigma_{r,i}(w_1, w_2, w_3, w_4)K_1(r, i)$, the word $\Sigma_{r,i}(w_1, w_2, w_3, w_4)K_1(r, i) \circ \tau$ also has the form $\Sigma_{r',i'}(w'_1, w'_2, w'_3, w'_4)K_1(r', i')$ for some $w'_1, w'_2, w'_3, w'_4, r', i'$. Now the statement of the lemma follows by $|h|$ applications of Lemma 3.5 a). $\square$

Lemma 3.7. *Assume a reduced word $w' = w'(a_1, \ldots, a_{\bar{m}})$ is freely equal to a product $wqrq^{-1}$, $r \in \mathcal{R}$. Then the words $\bar{\Sigma}(w)$ and $\bar{\Sigma}(w')$ are conjugate in $\mathcal{H}_0$. There is an annular diagram over $\mathcal{H}_0$ without θ-cells for this conjugacy, in which the contours p and p' can be connected by a path whose length is bounded by a linear function of $|w| + |w'| + |q|$.*

Proof. The proof is completely analogous to the proof of Lemma 3.6 $\square$

Lemma 3.8. *a) For every positive word w in the alphabet $\{a_1, \ldots, a_{\bar{m}}\}$ which is equal to 1 in $\bar{\mathcal{G}}$ the word $\Sigma(w)$ is a conjugate of the hub Σ in $\mathcal{H}_0$. The length of the conjugating word is a bounded by a recursive function in $|w|$. The corresponding annular diagram does not contain $\bar{\theta}$-cells.*

b) For every word w in the alphabet $\{a_1, \ldots, a_{\bar{m}}\}$ which is equal to 1 in $\bar{\mathcal{G}}$, the word $\bar{\Sigma}(w)$ is a conjugate of Σ in $\mathcal{H}_0$. The length of the conjugating word is a bounded by a recursive function in $|w|$. The corresponding annular diagram does not contain θ-cells.

Proof. It immediately follows from Lemmas 2.1, 3.6, 3.7 and the solvability of the word problem in $\bar{\mathcal{G}}$. $\square$

Lemma 3.9. *a) The map $a_1 \to a_1(P_1)$, $a_2 \to a_2(P_1)$, ..., $a_m \to a_m(P_1)$ extends to an injective homomorphism from $\mathcal{G}$ into $\mathcal{H}$.*

b) For every word w over $a_1, \ldots, a_m$ which is equal to 1 in $\mathcal{G}$, there exists a van Kampen diagram Δ over the presentation $\mathcal{R}$ of $\mathcal{H}$ with boundary label w and the number of cells bounded by a recursive function in $|w|$.

Proof. a) Let $w \in \mathcal{E}$. Then $w = \bar{w}$ (because we agreed to identify a_i with $\bar{a}_i$, $i = 1, ..., m$). By Lemma 3.8, both words $\Sigma(w)$ and $\bar{\Sigma}(w)$ are conjugate with $\Sigma = 1$ in $\mathcal{H}$. These words differ only by the subword $w(P_1)$ (i.e. $\Sigma(w)$ is obtained from $\bar{\Sigma}(w)$ by inserting $w(P_1)$). Hence $w(P_1) = 1$ in $\mathcal{H}$. Thus the map $a_i \to a_i(P_1)$ extends to a homomorphism from $\mathcal{G}$ to $\mathcal{H}$.

Since $\mathcal{G}$ is a subgroup of $\bar{\mathcal{G}}$, this homomorphism is injective by Lemma 3.1.

b) This statement immediately follows from Lemmas 2.1 and 3.8. □

3.3 Bands and annuli

Let $s^{\pm 1}$ be a letter, and S be a set of letters. We call a cell (edge) s-cell (s-edge) or S-cell (S-edge) provided its label contains a letter $s^{\pm 1}$ or a letter from $S \cup S^{-1}$, respectively. If a cell is both S_1-cell and S_2-cell, we call it a (S_1, S_2)-*cell*.

For example any cell corresponding to one of the relations (2.6), (2.7) is a $\Theta(\tau)$-cell.

For $z \in \{K, L, P, R\}$, we shall call a cell a z-*cell* or z_j-*cell* if it is a $z_j(r,i)$-cell for some j, r, i. Similarly we define $\bar{z}$-cells.

For simplicity we shall call an $\mathcal{A} \cup \bar{\mathcal{A}}$-edge, (letter, cell) as an a-edge (letter, cell), a $\Theta \cup \bar{\Theta}$-edge (letter, cell) as θ-edge(letter, cell), an $\mathcal{X}$-edge (letter, cell) as an x-edge (letter, cell), and $\mathcal{K} \cup \bar{\mathcal{K}}$-edge (letter, cell) k-edge (letter, cell). Thus we can use such notions as (a, x)-*cell*, (k, x)-*cell*, (θ, a)-*cell* and (θ, k)-*cell*.

As in [OlSa2], we enlarge the set of defining relations of the group $\mathcal{H}$ by adding some consequences of defining relations. Namely, by Lemma 3.9, we can include all cyclically reduced relations of the copy of the group $\mathcal{G}$ generated by the set $\{a_1(P_1), ..., a_m(P_1)\}$, i.e. all cyclically reduced words in $\{a_1(P_1), ..., a_m(P_1)\}$ which are equal to 1 in the copy of $\mathcal{G}$. These relations will be called $\mathcal{G}$-*relations*. The cells corresponding to $\mathcal{G}$-relations will be called $\mathcal{G}$-*cells*.

We denote by $\mathcal{H}_1$ the group given by all generators of $\mathcal{H}$, by all $\mathcal{G}$-relations, and by all defining relations of $\mathcal{H}$, except for the hub Σ.

The new, expanded, set of relations of $\mathcal{H}$ will be denoted by $\mathcal{R}'$.

It is convenient to turn $\mathcal{R}'$ into a *graded presentation*.

A hub is "higher" than (Θ, k)-cells corresponding to (2.6). (We also say that the *rank* of a hub is greater than the rank of a (Θ, k)-cell.) In turn, (Θ, k)-cells are "higher" than $(\bar{\Theta}, \bar{k})$-cells, which are "higher" than $\mathcal{G}$-cells which are "higher" than (Θ, a)-cells and $(\bar{\Theta}, \bar{a})$-cells which are "higher" than x-*cells* corresponding to relations (2.9), (2.8). The (Θ, k) and $(\bar{\Theta}, \bar{k})$-relations are also stratified: we require that relations (2.6) corresponding to rules τ from connecting steps (12), (34), (51) are "higher" than others.

If Δ and Δ' are diagrams over $\mathcal{R}'$ then we say that Δ has a higher *type* than Δ', if Δ has more hubs, or the numbers of hubs are the same, but Δ

has more cells which are next in the hierarchy, etc.

It is obvious, that in this way we define a partial quasi-order (a transitive reflexive relation) on the set of diagrams over $\mathcal{R}'$, and that this partial quasi-order satisfies the descending chain condition, so we can consider type as an inductive parameter.

A diagram Δ that has the smallest type among all diagrams with the given boundary label (boundary labels, in the annular case), then Δ is called a *minimal* diagram. The following lemma is obvious (see Figure 10).

Lemma 3.10. *Assume a diagram Δ over $\mathcal{H}$ has two cells Π and Π' of the same rank ρ such that words V and V' can be read on the boundary paths $\partial \Pi$ and $\partial \Pi'$ when going around the cells in opposite directions from vertices o and o'. Assume further that there is a path $p = o - o'$ in Δ with no self-intersections, $\phi(p) = U$, and there exists a diagram with boundary label $UV'U^{-1}V^{-1}$ containing at most one cell of rank ρ and some cells of smaller ranks. Then the number of cells of rank ρ in Δ can be decreased, while the boundary label (boundary labels, in the annular case) and numbers of cells of higher ranks are preserved. In particular Δ is not a minimal diagram.*

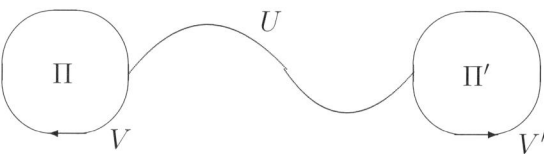

Figure 10.

We say that an annular diagram Γ over $\mathcal{H}_1$ is *compressible* if

(1) It contains no hubs, and there are no θ-edges on its boundary, and

(2) There exists another annular diagram Γ' which

   (a) has no hubs and has the same boundary labels as Γ,

   (b) the type of Γ' is smaller than the type of Γ, and these types differ already by the number of θ-cells, and

   (c) Γ' has no $(\bar{\Theta}, k)$-cells if Γ has no $(\bar{\Theta}, k)$-cells.

By definition, a diagram Δ is *not reduced* if either

(i) Δ contains a *reducible* pair of cells having a common edge $o - o'$ and boundary labels w and w^{-1} (read starting at the common vertex o) or

(ii) Δ contains a pair of (k, Θ)-cells like in Lemma 3.10 where $V' \equiv V$ and U is a word in $\mathcal{A}(P_1)$ which is equal to 1 modulo $\mathcal{G}$-relations, or

(iii) Δ contains a pair of $\mathcal{G}$-cells, and a connecting path p like in Lemma 3.10, where the label U of p is a word in $\theta(\tau, P_1)$-letters commuting with letters from $\mathcal{A}(P_1)$ by relations (2.7) or

(iv) Δ is an annular diagram and contains a compressible annular subdiagram surrounding the hole.

It is easy to see that every non-reduced diagram is not minimal. In case (iii), for example, the word U commutes with all $\mathcal{A}(P_1)$-letters by relations (2.7), hence the subdiagram consisting of the two $\mathcal{G}$-cells and the connecting path can be replaced by a subdiagram having at most one $\mathcal{G}$-cell and several commutativity (Θ, a)-cells. In case (iv), the annular subdiagram can be replaced by an annular diagram of a smaller type with the same boundary labels.

Of course, a diagram is said to be *reduced* if it is not non-reduced. Thus every minimal diagram is reduced (but the converse is not necessarily true).

As in our previous papers (see [OlSa2]), it is convenient to study properties of diagrams using bands. Let us fix a point o_Π inside every cell Π and some point o_Δ on the plane outside the diagram (or two points o and o' in each of two components of the complement of Δ on the plane, in the annular case). Similarly we fix points o_e inside every edge e of diagram Δ. In the case when e is an edge on $\partial(\Pi)$ (on $\partial(\Delta)$), we fix a simple Jordan curve $l(\Pi, e)$ (respectively $l(\Delta, e)$) that connects the vertex o_Π (o_Δ) with o_e and has no other common points (except for o_e) with the edges of the diagram Δ. We also require that $l(\Pi, e)$ and $l(\Pi, e')$ have the only common point o_Π for $e \neq e'$. The similar requirement concerns the lines $l(\Delta, e)$.

Let S be a set of letters and let Δ be a van Kampen diagram. Fix pairs of S-edges on the boundaries of some cells from Δ (we assume that each of these cells has exactly two S-edges).

Suppose that Δ contains a sequence of cells $(\pi_1, \ldots, \pi_n)$ such that for each $i = 1, \ldots, n-1$ the cells π_i and π_{i+1} have a common S-edge e_i and this edge belongs to the pair of S-edges fixed in π_i and π_{i+1}. Let e be the other S-edge ($\neq e_1$) of π_1, and let f be the other S-edge ($\neq e_{n-1}$) of π_n. Consider the line which is the union of the lines $\ell(\pi_i, e_i)$ and $\ell(\pi_{i+1}, e_i)$, $i = 1, \ldots, n-1$, together with $\ell(\pi_1, e)$ and $\ell(\pi_n, f)$. This broken line is called the *median* of this sequence of cells. Then our sequence of cells $(\pi_1, \ldots, \pi_n)$ with common S-edges $e_1, \ldots, e_{n-1}$, and distinguished edges e, f, is called an *S-band* (with start edge e and end edge f) if the median is an absolutely simple curve or an absolutely simple closed curve (see Fig. 11).

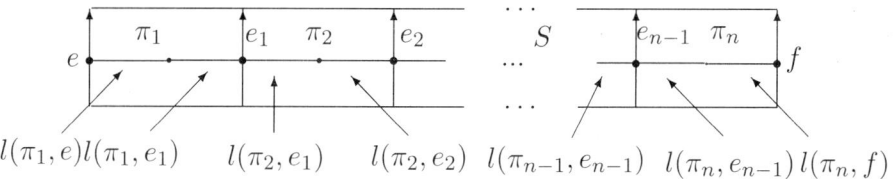

Fig. 11.

A band is said to be a *k-band* (θ-*band*, *a-band*) if S consists of k-letters (θ-letters, a-letters). Similarly, a band is an $a_i(z)$-band (a z_j-band) if S consists of one a-letter $a_i(z)$ (of $z_j(r,i)$- and $z_j(r,i)$-letters) for $z \in \tilde{\mathcal{K}}$.

Notice that an S-band may contain no cells. In this case the band is called *empty*.

We say that two bands *cross* if their medians cross. We say that a band is an *annulus* if its median is a closed curve. In this case the first and the last cells of the band coincide. (see Fig. 12a).

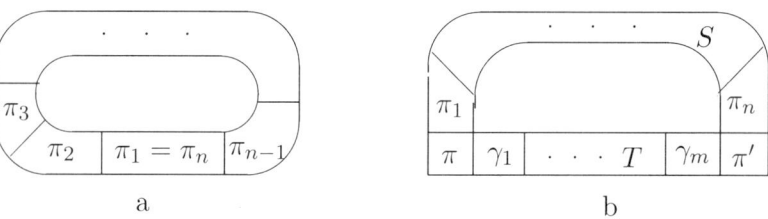

Fig. 12.

Let $\mathcal{B}$ be an S-band with common S-edges $e_1, e_2, \ldots, e_n$ which is not an annulus. Then the first cell has an S-edge e which forms a pair with e_1 and the last cell of $\mathcal{B}$ has an edge f which forms a pair with e_n. Then we shall say that e is the *start edge* of $\mathcal{B}$ and f is the *end edge* of $\mathcal{B}$. If p is a path in Δ then we shall say that a band starts (ends) on the path p if e (resp. f) belongs to p.

Let S and T be two disjoint sets of letters, let $(\pi, \pi_1, \ldots, \pi_n, \pi')$ be an S-band and let $(\pi, \gamma_1, \ldots, \gamma_m, \pi')$ is a T-band. Suppose that:

- the parts of the medians of these bands lying between their intersections, form a simple closed curve,

- on the boundary of π and on the boundary of π' the pairs of S-edges separate the pairs of T-edges,

- the start and end edges of these bands are not contained in the region bounded by the medians of the bands.

Then we say that these bands form an (S,T)-*annulus* and the simple curve formed by the medians of these bands is the *median* of this annulus (see Fig.

12b). Fig. 13 shows that a multiple intersection of an S-band and a T-band does not necessarily produce an (S, T)-annulus. Again, for simplicity, we say that this annulus is a (θ, k)-*annulus*, if S is a θ-band and T is a k-band. Similarly, the notion of a (θ, a)-*annulus* is defined.

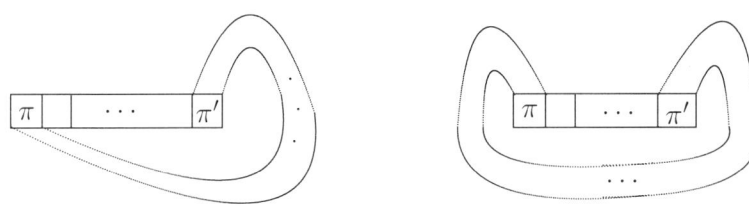

Fig. 13.

If ℓ is the median of an S-annulus or an (S, T)-annulus then the maximal subdiagram of Δ contained in the region bounded by ℓ is called the *inner diagram* of the annulus.

The union of cells of an S-band $\mathcal{B}$ forms a subdiagram. The reduced boundary of this diagram, which we shall call the *boundary of the band*, has the form $e^{\pm 1} p f^{\pm 1} q^{-1}$ (recall that we trace boundaries of diagrams clockwise). Then we say that p is the *top path* of $\mathcal{B}$, denoted by $\mathbf{top}(\mathcal{B})$, and q is the *bottom path* of $\mathcal{B}$, denoted by $\mathbf{bot}(\mathcal{B})$.

We shall call an S-band *maximal* if it is not contained in any other S-band. If an S-band $\mathcal{W}$ starts on the contour of a cell π, does not contain π and is not contained in any other S-band with these properties then we call $\mathcal{W}$ a *maximal S-band starting on the contour of π*.

We now list all types of bands in diagrams over the presentation $\mathcal{R}'$, that will be considered further.

1. By definition, any Θ-cell can be included in a Θ-band (Θ-annulus) of a diagram Δ over $\mathcal{R}'$. The cells in these bands correspond to the relations (2.6), (2.7). Similarly $\bar{\Theta}$-bands and $\bar{\Theta}$-annulus are defined.

It is easy to see from (2.6) and (2.7) that if a θ-band contains a $\theta(\tau, z)$-cell, for some $\tau \in \mathcal{S} \cup \bar{\mathcal{S}}$, then all cells in that band correspond to the same rule τ. Thus we can talk about $\Theta(\tau)$-bands and $\bar{\Theta}(\tau)$-bands. Since every θ-cell has two θ-edges, any maximal θ-band, which is not an annulus, must start on a boundary edge of the diagram.

2. By definition, an $a_j(z)$-band, where $z \in \tilde{\mathcal{K}}$, $a_j \in \mathcal{A} \cup \bar{\mathcal{A}}$, is constructed of $(a_j(z), \theta)$-cells and $(a_j(z), x)$-cells (see relations (2.7), (2.8)).

These bands can start and end on the boundary of a diagram, on (θ, k)-cells and on $\mathcal{G}$-cells. Thus, an a-band is a $a_j(z)$-band for some j, z.

3. For every $k \in \tilde{\mathcal{K}}$, a k-band is constructed of (z, θ)- and (z, x)-cells where $z \in \mathcal{K} \cup \tilde{\mathcal{K}}$ and z projects onto k (see 2.3). These bands can start and end on the boundary of the diagram or on hubs.

Notice that the reducibility of the top and the bottom labels of a θ-band can be easily achieved without changing the boundary or the type of the diagram (see [SBR]).

Indeed, if the label of the path **bot**($\mathcal{B}$) is not reduced then it contains a subpath $e_1 e_2$ where e_1, e_2 are edges labelled by a and a^{-1} respectively, where a is a letter. These two edges must belong to two different cells π_1 and π_2 of $\mathcal{B}$ as in the diagram on the left in the Figure 14.

Then we fold e_1 and e_2, producing a new edge e labelled by a and we introduce a new edge f with label a which has a common end vertex with e so that cells π_1 and π_2 have a common edge e. The other cells which were attached to e_1 and e_2 will be attached to the edge f. The bottom path of the θ-band $\mathcal{B}$ in the new diagram is shorter (by two edges) than the bottom path of $\mathcal{B}$ in Δ. The bottom paths of the other θ-bands are not affected by this operation. Thus after a finite number of such operations we get a diagram with the same boundary label as Δ, in which all bands $\mathcal{B}$ have top and bottom paths with reduced labels.

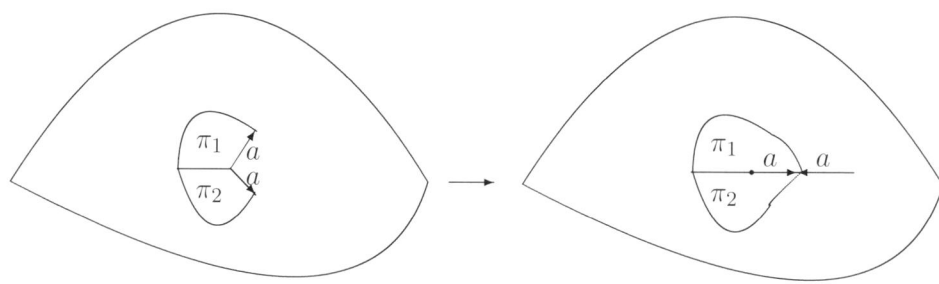

Figure 14.

Thus we will change the definition of reduced diagrams by demanding in addition that the top and bottom labels of any θ-band in a reduced diagram are reduced words.

3.4 Forbidden annuli

Similarly to [OlSa2], [OlSa1] the absence of annuli of various kinds in reduced diagrams without hubs is important in our present considerations. The following lemma is similar to Lemma 3.1 of [OlSa1].

Lemma 3.11. *Let Δ be a reduced (in particular, a minimal) van Kampen diagram over the group $\mathcal{H}_1$. Then Δ has no*
  (1) θ-annuli;
  (2) k-annuli;

(3) a-annuli;
(4) (θ, k)-annuli;
(5) (θ, a)-annuli;

Proof. We prove the lemma by contradiction. Let us choose a minimal annulus $\mathcal{T}$ in Δ among all counterexamples to any of the assertions (1) - (5). Here "minimal" means that the inner subdiagram of the annulus is of the smallest possible type.

Let $\mathcal{T}$ be a θ-annulus. Suppose it contains a (θ, k)-cell. Notice that in every (θ, k)-cell the pair of θ-edges separates cyclically the pair of k-edges. Hence both top and bottom sides of $\mathcal{T}$ contain k-edges. Thus the inside subdiagram of $\mathcal{T}$ contains a k-band starting on the top or bottom of $\mathcal{T}$. Since Δ does not have hubs, there exists a maximal k-band $\mathcal{L}$ starting and ending on the contour of the inside diagram. This gives rise to a (θ, k)-annulus with a smaller inside subdiagram than $\mathcal{T}$ which contradicts the choice of $\mathcal{T}$ (as the smallest counterexample) and property (4) of the lemma. Hence all cells in $\mathcal{T}$ are (θ, a)-cells corresponding to the auxiliary relations (2.7).

Similarly, the inside subdiagram Δ' of $\mathcal{T}$ contains no k-edges. Hence the only possible cells in Δ' where an a-band can end would be $\mathcal{G}$-cells (whose contours consist of $a_i(P_1)$-edges). Hence if Δ' contains an $a_i(z)$-edge, and $z \neq P_1$, the maximal $a_i(z)$-band of Δ' containing that cell must start and end on the boundary of Δ'. This gives rise to a (θ, a)-annulus ruled out by property (5) of the lemma. Thus all a-edges in Δ' are $\mathcal{A}(P_1)$-edges. Hence all cells in $\mathcal{T}$ are $(\Theta, \mathcal{A}(P_1))$-cells corresponding to relations (2.7). Notice that all these relations are commutativity relations because $\alpha_{\tau^{-1}}(a_i(P_1)) = \alpha_\tau(a_i(P_1)) = a_i(P_1)$. Hence the labels of the top and the bottom paths of $\mathcal{T}$ coincide which contradicts the assumption that Δ is reduced (see part (iv) of the definition of non-reduced diagrams).

Now, let $\mathcal{T}$ be a (θ, k)-annulus, $\mathcal{T}_\theta$ its θ-band, and $\mathcal{T}_k$ its k-band. We denote by π and π' their common cells, i.e. the *corner cells* of the annulus $\mathcal{T}$. The inner diagram Γ of T is bounded by a path $p_\theta p_k$ where p_θ is a subpath of $\partial(\mathcal{T}_\theta)$ and p_k is a subpath of $\partial(\mathcal{T}_k)$.

Property (4), which holds for smaller annuli, ensures that there are no θ-edges in p_k and there are no k-edges in p_θ. Hence every non-corner cell of $\mathcal{T}_k$ is a cell corresponding to relations (2.9). There exist no such cells if $\mathcal{T}_k$ is a P_j- or R_j-band for any j. Then π and π' must have a common k-edge. Moreover, π is a mirror copy of π', since these cells belong to the same θ-band $\mathcal{T}_\theta$ as well. But this contradicts the assumption that the diagram Δ is reduced. Hence we may assume that $\mathcal{T}_k$ is neither P- nor R-band.

Since every non-corner cell of $\mathcal{T}_k$ corresponds to (2.9), we have $\phi(p_k) \equiv U$ where U is a word in $\mathcal{X}$ (see relations (2.9). If p_k contains a-edges of the corner cells, then we just make the path p_k shorter and make p_θ longer, respectively, keeping the boundary path of Γ equal to $p_\theta p_k$.

Again, by (4), for smaller diagrams, every non-corner cell of $\mathfrak{T}_\theta$ is a (θ, a)-cell, and the forms of relations (2.7), (2.6) show that $\phi(p_\theta) \equiv \alpha_{\tau \pm 1}(V)$ for some V in a-letters. The word V has no letters from $\mathcal{A}(P_1)$ because π and π' are neither P_1- nor R_1-cells.

The equality $U\alpha_{\tau \pm 1}(V) = 1$ given by Γ holds over the group $\mathcal{H}_2$ given by all the x-, k-, and a-generators, and x-relations (2.8) and (2.9) (but not (2.7)). Indeed, by statements (1) - (5) for smaller annuli, Γ may contain only cells corresponding to relations of $\mathcal{H}_2$ and $\mathcal{G}$-cells. If we collapse all $\mathcal{A}(P_1)$-edges in Γ, we obtain a diagram over $\mathcal{H}_2$ with the same boundary label (since $\partial(\Gamma)$ does not contain $\mathcal{A}(P_1)$-edges). Thus we can assume that Γ does not contain $\mathcal{G}$-cells.

The group $\mathcal{H}_2$ is a (multiple) HNN-extension of the free group generated by x-letters where non-x-letters are stable. Hence V and $\alpha_{\tau \pm 1}(V)$ are freely trivial words. Then so is the word U.

Since there are no reducible pairs of (k, x)-cells in $\mathfrak{T}_k$, and $U = 1$ in the free group, we see that π and π' must have a common k-edge, and we come to a contradiction as above.

Similar arguments work in other cases (see lemma 3.1 in [OlSa1] and lemma 6.1 in [Ol2] for details). □

3.5 Surgeries involving $\mathcal{G}$-cells

As in our previous papers, we now describe certain surgeries which can be applied to a diagram to reduce its type.

Lemma 3.12. *(1) (**Moving a $\mathcal{G}$-cell along an a-band.**) Assume Δ is a van Kampen diagram formed by a $\mathcal{G}$-cell Π and an $\mathcal{A}(P_1)$-band $\mathfrak{T}$ attached to Π along an $\mathcal{A}(P_1)$-edge e. Let $e^{-1}p$ and $eqe'q'$ be clockwise contours of Π and $\mathfrak{T}$, respectively, where e and e' are a-edges. Then there is a reduced diagram Δ' with contour $\tilde{p}\tilde{q}\tilde{e}'\tilde{q}'$, where the labels of $\tilde{p}, \tilde{q}, \tilde{e}', \tilde{q}'$ coincide with the labels of p, q, e', q', respectively, such that Δ' is formed by a copy Π' of the $\mathcal{G}$-cell Π, attached to $\bar{e}'$, and by several $(\Theta, \mathcal{A}(P_1))$-cells.*

*(2) (**Merging two $\mathcal{G}$-cells connected by an a-band.**) Let Δ be a van Kampen diagram consisting of two distinct $\mathcal{G}$-cells Π_1, Π_2 and an $a_i(P_1)$-band $\mathfrak{T}$ connecting them. Then there is a reduced diagram Δ' with the same boundary label, which has at most one $\mathcal{A}$-cell and several $(\Theta, \mathcal{A}(P_1))$-cells; and therefore the diagram Δ is not reduced.*

(3) There does not exist a reduced van Kampen diagram over $\mathcal{H}_1$ which contains a $\mathcal{G}$-cell Π and an $\mathcal{A}$-band $\mathfrak{T}$ that starts and ends on Π.

Proof. (1) Consider the cell π_1 of $\mathfrak{T}$ containing e in the boundary. By relations (2.7), the Θ-letter θ labelling two edges of π_1, commutes with all $\mathcal{A}(P_1)$-letters in $\mathcal{H}_1$. Hence Π can be surrounded by a θ-annulus (i.e. there exists a Θ-annulus containing π such that the boundary label of the internal subdiagram of this annulus coincides with the boundary label of Π, so we

THE CONJUGACY PROBLEM AND HIGMAN EMBEDDINGS 35

can glue the annulus around Π). Next we can glue the resulting diagram inside a Θ-annuli starting containing the second cell π_2 of $\mathcal{T}$, and so on. Notice that the boundary labels of all diagrams constructed this way coincide with the boundary label of Π because all cells in the surrounding annuli are commutativity cells.

Thus, we obtain a diagram Δ_0 with the same boundary label as Π, such that Δ_0 contains attached to e'. This subdiagram can be replaced by a copy Π' of Π. Finally, to restore the boundary label of Δ, we can attach a subdiagram Γ to Π', where Γ is the mirror copy of $\Delta_0 \backslash \Pi$. The resulting diagram is what we need. Figure 15 illustrates the procedure used in this proof.

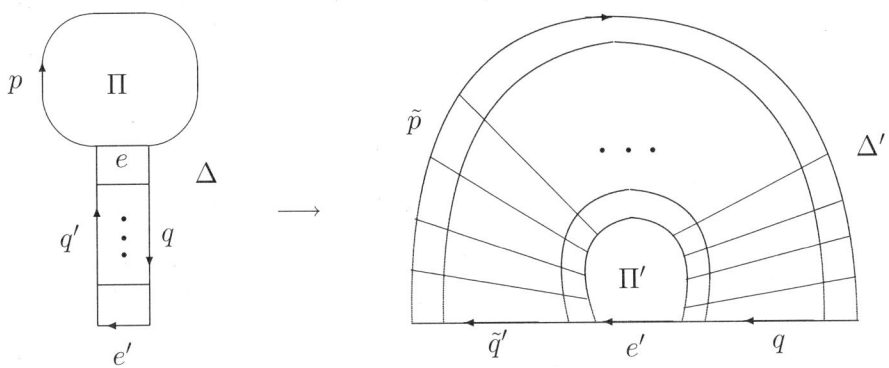

Figure 15.

(2) Indeed, the label of the top path of the connecting a-band commutes with all letters in $\mathcal{A}(P_1)$. Hence we can apply Lemma 3.10 (see also part (iii) of the definition of non-reduced diagrams).

(3) Arguing by contradiction, consider the subdiagram Γ bounded by $\partial(Pi)$ and $\mathcal{T}$. It follows from Lemma 3.11, that Γ has no cells except for $\mathcal{G}$-cells, and that $|\mathcal{T}| = 0$. If Γ contains a $\mathcal{G}$-cell, it can be merged with Π. If Γ does not contain cells, we obtain a contradiction with the assumption that the boundary label of Π is cyclically reduced. $\square$

3.6 Shifting indexes

The following lemma utilizes the fact that our $\mathcal{S}$-machines work the same way in all parts of words $\Sigma_{r,i}(w_1, w_2, w_3, w_4)$ between two consecutive K_j-letters.

Recall that every letter in the list of generators of $\mathcal{H}$ has the form $z(r, i)$ or $a_s(z)$ or $\theta(\tau, z)$, or $x(a_s(z), \tau)$ or is a "bar"-brother of such a letter, where $z \in \{\bar{L}_j, L_j, P_j, R_j\}$, $j = 1, ..., N$, $r \in \bar{\mathcal{E}}, i = 1, 2, 3, 4, 5$, $1 \le s \le \bar{m}$. That j will be called the *index* of the letter. We say that a word W has index j if all letters in W have index j. If W has index j and $j' = 1, ..., N$, then $\varepsilon_{j'}(W)$ denotes the word W where the index j is replaced by j' in all letters. The word $\bar{\varepsilon}_{j'}(W)$ coincides with $\varepsilon_{j'}(W)$ if $j' \neq 1$ and is obtained from $\varepsilon_{j'}(W)$ by removing all $\mathcal{A}$-letters if $j' = 1$.

Lemma 3.13. *Suppose an equality $W = 1$ holds in the group $\mathcal{H}_1$ where W has index $j \neq 1$ and has no $\bar{\Theta}$-letters, K_l-letters and $\bar{K}_l$-letters ($l = 1, ..., N$). Then $\varepsilon_{j'}(W) = 1$ in $\mathcal{H}_1$ for any $j' = 1, ..., N$.*

Proof. Let Δ be a minimal van Kampen diagram with boundary label W. By Lemma 3.11, Δ has no $\bar{\Theta}$-edges since its boundary has no $\bar{\Theta}$-edges.

Similarly, the diagram Δ does not contain K_l-edges and $\bar{K}_l$-edges for any $l = 1, ..., N$, and if e is a z-edge in Δ, $z \in \{L_{j_0}, P_{j_0}, R_{j_0}\}$, then $j_0 = j$. Hence if a maximal $\mathcal{A}(z)$-band in Δ terminates on a Θ-cell inside Δ, then $z \in \{L_j, P_j, R_j\}$.

Since all a-edges of the boundary of Δ belong to $\mathcal{A}(z) \cup \bar{\mathcal{A}}(z)$, $z \in \{L_j, P_j, R_j\}$, all a-edges inside Δ must have labels from $\mathcal{A}(z) \cup \bar{\mathcal{A}}(z) \cup \mathcal{A}(P_1) \cup \bar{\mathcal{A}}(P_1)$. If Δ contains $\mathcal{A}(P_1) \cup \bar{\mathcal{A}}(P_1)$-edges, it must contain $\mathcal{G}$-cells. By Lemma 3.12 (3) any a-band starting on any $\mathcal{G}$-cell must end on another $\mathcal{G}$-cell which by Lemma 3.12 (2), (3) contradicts minimality of Δ. Hence Δ does not contain $\mathcal{G}$-cells, and all a-edges of Δ are $\mathcal{A}(z) \cup \bar{\mathcal{A}}(z)$-edges, $z \in \{L_j, P_j, R_j\}$.

Since every x-cell contains a $\mathcal{A}$-letter or a z-letter, $z \in \mathcal{K}$ and the indexes of all letters in an a-cell (resp. a z-cell, $z \neq K_l$), are the same, all labels of x-edges in Δ have index j.

Thus the indexes of the labels of all edges in Δ are equal to j. When we replace the index j by j' in all letters of Δ, we obtain a new diagram Δ' with boundary label $\varepsilon_{j'}(W)$. Since the $\varepsilon_{j'}$ takes relations of $\mathcal{H}_1$ without K_l-letters, $\bar{K}_l$-letters and $\bar{\Theta}$-letters to relations of $\mathcal{H}_1$, the diagram Δ' is a van Kampen diagram over $\mathcal{H}_1$, so the equality $\varepsilon_{j'}(W) = 1$ holds in $\mathcal{H}_1$ as required. □

Lemma 3.14. *Suppose an equality $W = 1$ holds in the group $\mathcal{H}_1$, where the word W has index j for some $j \neq 1$ and does not contain Θ-letters, K_l-letters, and $\bar{K}_l$-letters for any $l = 1, ..., N$. Assume that a reduced van Kampen diagram over $\mathcal{H}_1$ for the equality $W = 1$ contains no $(\mathcal{A}, \mathcal{X})$-cells. Then $\bar{\varepsilon}_{j'}(W) = 1$ in $\mathcal{H}_1$ for any $j' = 1, ..., N$.*

Proof. By Lemma 3.11, Δ has no $\bar{K}_l$-edges and K_l-edges, since there are no such edges on the boundary of Δ. Therefore the indexes of all labels of edges on the contour of every cell in Δ are the same. Similarly, it does not contain Θ-edges, and the labels of all $\bar{\Theta}$-edges in Δ have index j.

Since every a-band in Δ ends either on the boundary of Δ or on a $\bar{\Theta}$-cell, all a-edges have index j and they are $\bar{\mathcal{A}}$-edges. Similarly all x-edges in Δ (if any) have index j as well.

Now if $j' \neq 1$, when we replace the index of every label in Δ by j', we get a van Kampen diagram over $\mathcal{H}_1$ with boundary label $\bar{\varepsilon}_{j'}(W)$ as required.

Let $j' = 1$. It is easy to see that when we apply $\bar{\varepsilon}_1$ to every defining relation of $\mathcal{H}_1$ occurring in Δ, we get a relation of $\mathcal{H}_1$ again. Hence if we apply $\bar{\varepsilon}_1$ to all labels of Δ, we obtain a van Kampen diagram over $\mathcal{H}_1$ with boundary label $\bar{\varepsilon}_1(W)$, so $\bar{\varepsilon}_1(W) = 1$ in $\mathcal{H}_1$ as required. □

4 The group $\mathcal{H}_2$

Recall that $\mathcal{H}_2$ is the auxiliary group given by all the x-, k-, and a-generators, and by x-relations (2.8) and (2.9) (but not (2.7)).

Lemma 4.1. *The word problem is solvable for the group $\mathcal{H}_2$.*

Proof. The group $\mathcal{H}_2$ is an HNN-extension of the free group generated by x-letters where non-x-letters are stable. The associated subgroups are either subgroups generated by $\mathcal{X}(z)$, $z \in \tilde{\mathcal{K}}$, or subgroups generated by fourth powers of elements from $\mathcal{X}(z)$, $z \in \tilde{\mathcal{K}}$. Hence the membership problem is decidable for all associated subgroups. So the claim of the lemma follows from Britton's lemma [LS]. □

We say that a non-empty cyclically reduced word w is *uniform* if it is written in $\mathcal{X}(z)$ for some $z \in \tilde{\mathcal{K}}$, $z \neq P_j$. Two uniform words w_1 and w_2 are said to be *related* if one of them can be obtained from another one, by a sequence of substitutions of the following two forms, or their inverses:

(1) Substitution $x \to x^4$ applied to every letter of the word;

(2) Every letter in the word is replaced by the corresponding letter in $\mathcal{X}(z_-)$ if $z, z_- \neq P_j$.

It is obvious that the problem of whether two words in $\mathcal{X}$ are related is decidable.

Lemma 4.2. *Let $\mathcal{B}$ be an a- or a k-band in a diagram over $\mathcal{H}_2$. Then the label of the top (bottom) path of $\mathcal{B}$ is a uniform word. The labels of the top and the bottom paths are related.*

Proof. This immediately follows from relations (2.8) and (2.9). □

Lemma 4.3. *Two non-empty cyclically reduced words w_1 and w_2 in $\mathcal{X}$ are conjugate in $\mathcal{H}_2$ if and only if one of them is a cyclic permutation of another one, or a cyclic shift of w_1 and a cyclic shift of w_2 are related uniform words.*

Proof. The condition is sufficient for the conjugacy of w_1 and w_2 this follows from the conjugacy relations (2.8) and (2.9) in $\mathcal{H}_2$.

Now suppose that w_1 and w_2 are conjugate in $\mathcal{H}_2$. Consider a minimal annular diagram Δ for the conjugacy of w_1 and w_2, w_1 (resp. w_2) is the label of the inner (outer) boundary of Δ. We may assume that w_2 is not a cyclic shift of w_1, and so the diagram Δ contains cells. These cells belong to concentric a- and k-annuli $\mathcal{B}_1, \ldots, \mathcal{B}_d$, counted from the inner boundary to the outer boundary of Δ.

Therefore by Lemma 4.2 each of the words w_1, w_2 must be uniform, being a side label of the annuli $\mathcal{B}_1$ or of $\mathcal{B}_d$. Also by Lemma 4.2, a cyclic shift of w_1 and w_2 are related. $\square$

Lemma 4.4. *The conjugacy problem is solvable in the group $\mathcal{H}_2$.*

Proof. It suffices to find an effective upper bound for the number of cells in an annular diagram Δ over $\mathcal{H}_2$, in terms of the sum of lengths of the inner and outer boundaries p_1, p_2 of Δ.

By Lemma 4.1, we may assume that the boundary labels $w_1 = \phi(p_1)$ and $w_2 = \phi(p_2)$ of Δ are non-empty words which have no pinches over the HNN-extension $\mathcal{H}_2$. This implies that there is no a- or k-band in Δ, with both ends lying on p_1 (resp. on p_2). Indeed, otherwise there would exist a band $\mathcal{B}$ with ends e_1, e_2 lying, say, on p_1, such that, the bottom or the top of $\mathcal{B}$ is a subpath of p_1 (since each cell of Δ belongs to a maximal a- or k-band); then label of this side of $\mathcal{B}$ would be a pinch in w_1.

Obviously, Lemma 4.3 solves the conjugacy problem for words in $\mathcal{X}$. So we can assume that one of the words w_1 or w_2 contains a non-$\mathcal{X}$-letter. Hence every maximal a- and k-band of Δ connects p_1 with p_2.

Let us enumerate these "radial" bands cyclically (counter-clockwise): $\mathcal{B}_1, \ldots, \mathcal{B}_d$. For each $t = 1, \ldots, d$, the boundary of $\mathcal{B}_t$ has the form

$$e_t \mathbf{top}(\mathcal{B}_t) f_t^{-1} \mathbf{bot}(\mathcal{B}_t)^{-1},$$

where the start edge e_t belongs to p_1, the end edge f_t belongs to p_2.

Notice that the labels of the paths $\mathbf{top}(\mathcal{B}_t)$ and $\mathbf{bot}(\mathcal{B}_t)$ are reduced uniform words in $\mathcal{X}$ by Lemma 4.2. Notice also that $\phi(\mathbf{top}(\mathcal{B}_t))$ coincides with $\phi(\mathbf{bot}(\mathcal{B}_{t+1}))$ (indices are taken modulo d) except for some prefixes and suffixes whose total length is bounded from above by a constant $c_0 = |w_1| + |w_2|$. Thus, in particular, $||\mathbf{bot}(\mathcal{B}_{t+1})| - |\mathbf{top}(\mathcal{B}_t)|| \leq c_0$.

Relations (2.8) and (2.9) show that for every $t = 1, \ldots, d$ we have

$$|\mathbf{top}(\mathcal{B}_t)| = c_t |\mathbf{bot}(\mathcal{B}_t)|$$

where $c_t \in \{1/4, 1, 4\}$. If $c_t = 1/4$, we call the band $\mathcal{B}_t$ *contracting*, if $c_t = 4$, we call it *expanding*.

Therefore, $|\mathbf{top}(\mathcal{B}_{t+1})| = c_{t+1} |\mathbf{top}(\mathcal{B}_t)| + c'_{t+1}$ for some $c'_{t+1} \leq c_0$. Hence

$$|\mathbf{top}(\mathcal{B}_{t+2})| = c_{t+2} c_{t+1} |\mathbf{top}(\mathcal{B}_t)| + c_{t+2} c'_{t+1} + c'_{t+2}.$$

Continuing this way, we finally obtain

$$|\mathbf{top}(\mathcal{B}_1)| \leq 4^\sigma |\mathbf{top}(\mathcal{B}_1)| + c$$

where σ is the number of expanding bands minus the number of contracting bands among $\mathcal{B}_1, ..., \mathcal{B}_d$, and c is some constant which is bounded by $c_0(1 + 4 + ... + 4^{d-1}) < c_0 4^d/2$.

If $\sigma \neq 0$, then $|\mathbf{top}(\mathcal{B}_1)| \leq c_0 4^d/2$. Hence $\mathbf{top}(\mathcal{B}_1)$ has a recursively bounded length (in terms of $|w_1|+|w_2|$). Cutting Δ along $\mathbf{top}(\mathcal{B}_1)$, we obtain an (ordinary, simply connected) van Kampen diagram Γ with perimeter recursively bounded in terms of $|w_1| + |w_2|$. Clearly, Γ is a minimal diagram over the presentation of $\mathcal{H}_2$ (otherwise there would be another diagram Γ' with the same boundary label and a smaller type; by gluing together two copies of $\mathbf{top}(\mathcal{B}_1)$ on the boundary of Γ', we obtain a annular diagrams Δ' with boundary labels w_1, w_2, whose type is smaller than that of Δ, a contradiction). By Lemma 4.1 the number of cells in the diagram Γ is recursively bounded in terms of $|w_1| + |w_2|$. Since Γ and Δ have the same number of cells, the number of cells in Δ is recursively bounded.

Thus we may assume that $\sigma = 0$.

Let us choose an arbitrary vertex $o = o_1$ on the path $\mathbf{top}(\mathcal{B}_1)$ far enough from the initial and terminal vertices of $\mathcal{B}_1$. Here "enough" means "farther than some constant $C > c_0$ whose value will become clear later".

Denote by $l_1 > C$ the distance from o_1 to the initial vertex of $\mathbf{top}(\mathcal{B}_1)$. Since $l_1 > c_0$, the vertex o_1 can be connected with a vertex o_2 of the path $\mathbf{top}(\mathcal{B}_2)$ by a path of the length at most 4 (going at most three edges along $\mathbf{top}(\mathcal{B}_1)$ toward its initial vertex and then across the band $\mathcal{B}_2$ along a connecting edge. Let l_2 be the distance from o_2 to the initial vertex of $\mathbf{top}(\mathcal{B}_2)$. It is clear that $l_2 = c_2 l_1 + c_2''$ for some recursively bounded constant c_2'', $c_2 \in \{4, 1/4, 1\}$. We can assume that C is big enough so that $l_2 > c_0$, so we can continue and find a vertex o_3 on $\mathbf{top}(\mathcal{B}_3)$ connected with o_2 by a path of length at most 4. Thus we can construct a sequence of points $o_i \in \mathbf{top}(\mathcal{B}_i)$, $i = 1, ..., d+1$ (where $\mathcal{B}_{d+1} = \mathcal{B}_1$) and a path p connecting these points such that $|p| \leq 4d$, and the distance $|l_{d+1} - l_1|$ from o_{d+1} to o_1 on $\mathbf{top}(\mathcal{B}_1)$ is equal $4^\sigma l_1 + C'$ for some recursively bounded constant C' (see Figure 16). Let p' be the loop obtained by concatenation of p and a portion of $\mathbf{top}(\mathcal{B}_1)$ between o_1 and o_{d+1}. Since $\sigma = 0$, the length of p' is bounded by a constant C''.

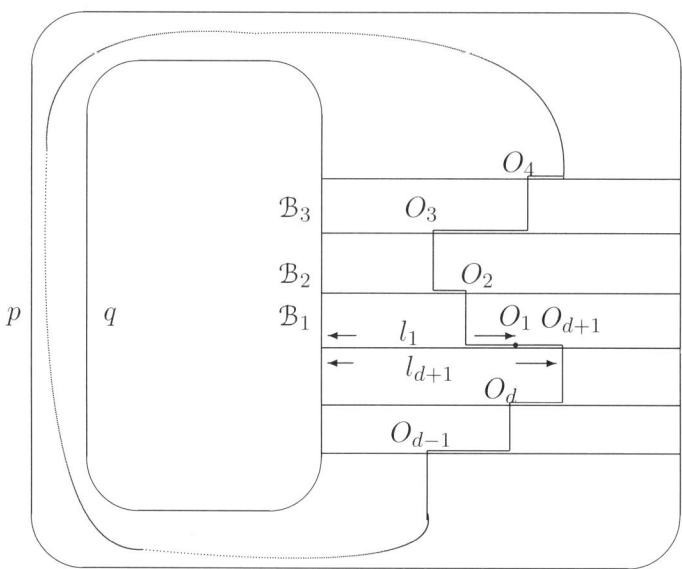

Figure 16.

If the length of **top**($\mathcal{B}_1$) is very large, then by choosing different starting points o_1 on **top**($\mathcal{B}_1$), we can built very many loops p' whose lengths are bounded by C'''. Notice that since the process of building the loop p' starting at o_1 is deterministic, if two of these loops share a common vertex, they will share all the vertices coming after that one. Hence if we take initial vertices of these loops sufficiently far apart (say, farther than $2C'''$), these loops won't intersect. The number of different labels of such loops is recursively bounded since their lengths are bounded. Hence if |**top**($\mathcal{B}_1$)| were bigger than that bound, two of the constructed paths would have the same labels. But this would contradict the minimality of Δ (in fact Δ would not be even reduced). Therefore |**top**($\mathcal{B}_1$)| is recursively bounded, and so we again found a cut of Δ which has a recursively bounded length and turns Δ into a simply connected van Kampen diagram with recursively bounded perimeter. □

Lemma 4.5. *Let Δ be a reduced diagram over $\mathcal{H}_1$ with boundary*

$$ep_1 e' q_1 (fp_2 f')^{-1} q_2^{-1},$$

and the following conditions hold:

(1) e, e', f, f' *are Θ-edges,*

(2) $\phi(p_1), \phi(p_2)$ *are reduced words in $\mathcal{X}$,*

(3) $q_1 = $ **top**($\mathcal{T}_1$), $q_2 = $ **bot**($\mathcal{T}_2$) *for some Θ-bands $\mathcal{T}_1, \mathcal{T}_2$ without k- and $a(P_j)$-cells,*

(4) The bands $\mathcal{T}_1, \mathcal{T}_2$ are decomposed as $\mathcal{T}_1 = \mathcal{T}'_1\mathcal{T}''_1$, $\mathcal{T}_2 = \mathcal{T}'_2\mathcal{T}''_2$ such that the words $\phi(\mathbf{bot}(\mathcal{T}'_1))_a$, $\phi(\mathbf{bot}(\mathcal{T}''_1))_a^{-1}$, $\phi(\mathbf{top}(\mathcal{T}'_2))_a$ and $\phi(\mathbf{top}(\mathcal{T}''_2))_a^{-1}$ are positive,

(5) The diagram $\Delta \setminus (\mathcal{T}_1 \cup \mathcal{T}_2)$ contains no θ-edges.

Then

(i) the length of every a-band connecting $\mathbf{bot}(\mathcal{T}'_1)$ and $\mathbf{top}(\mathcal{T}'_2)$ is between $\frac{|p_1|}{6^{|\mathcal{T}'_1|}} - 2$ and $|p_1|$,

(ii) the length of an a-band, connecting $\mathbf{bot}(\mathcal{T}''_1)$ and $\mathbf{top}(\mathcal{T}''_2)$ is between $\frac{|p_2|}{6^{|\mathcal{T}''_1|}} - 2$ and $|p_2|$.

(iii) $|\mathcal{T}'_1| \leq 2|p_1| + 2$.

Proof. Parts (i) and (ii) of the lemma are symmetric so it is enough to prove part (i). Denote by q'_1 (by q'_2) the bottom (the top) of $\mathcal{T}'_1$ (of $\mathcal{T}''_2$). Let $\mathcal{C}_1, \mathcal{C}_2$ be two neighbor maximal a-bands in Δ corresponding to two consecutive a-letters in the words $\phi(q'_1)_a$ and $\phi(q'_2)_a$. Denote by U_1 and U'_1 (by U_2 and U'_2) the top and bottom labels of $\mathcal{C}_1$ (of $\mathcal{C}_2$), respectively. By the positiveness condition (4) and the definition of α_τ, we have that U_1 is freely equal to $x^{\pm 1}U'_2(x')^{\pm 1}$ for some x-letters x, x', and U'_2 is obtained from U_2 by replacing all letters by their fourth powers (see relations (2.8)). Hence the length of $\mathcal{C}_1$ is strictly greater than the length of $\mathcal{C}_2$ if $|U_2| \neq 0$. If U_2 is empty, then $x \neq x'$ because otherwise the θ-cells of $\mathcal{T}_1$ and $\mathcal{T}_2$, connected by $\mathcal{C}_2$, correspond to the same rule τ (defined by x) and form a reducible pair of cells. Hence, in any case $|\mathcal{C}_1| > |\mathcal{C}_2|$. Since the length of the first maximal a-band in Δ (counting from p_1) is at most $|p_1|$, we obtain the upper estimate in part (i). Also we have from above considerations, that $|U_1| \leq 6|U_2|$ if $|U_2| \neq 0$, and this immediately implies the lower estimate in parts (i) of the lemma.

Part (iii) is obtained as follows. The number of maximal a-bands starting on q'_1 cannot exceed $|p_1|+1$ because the length of the next a-band is smaller than the length of the previous one. Also the length of q'_1 does not exceed twice the number of a-edges on q'_1. Therefore $|q'_1| \leq 2|p_1| + 2$. $\square$

Lemma 4.6. *Let Δ be a diagram over $\mathcal{H}_2$ with reduced boundary $pq_1q_2q_3$, and for some $\tau \in \mathcal{S}^+$, $\phi(q_1) = \alpha_{\tau^{\pm 1}}(u)$, $\phi(q_3) = \alpha_{\tau'^{\pm 1}}(v)$ for some reduced words u, v in $\mathcal{A}$, and $\phi(q_2)$ is a reduced word in $\mathcal{X}$. Suppose further that every a-band that starts on q_1 ends on p. Then the number of cells in Δ and the length of $|q_1q_2q_3|$ are recursively bounded in terms of $|p|$.*

Proof. We can assume that there are no a- or k-bands starting and ending on p. Otherwise we could cut off one such band whose side is a subpath of p, producing a diagram of smaller type satisfying the conditions of the lemma (with side p replaced by a path p' whose length is at most $4|p|$).

Thus by Lemma 3.11 we can assume that every cell in Δ belongs to an a-band starting on $q_1 \cup q_3$ and ending on p. The number l of such bands is at most $|p|$. Let us number these bands: $\mathcal{C}_1, \mathcal{C}_2, ..., \mathcal{C}_{l_1}, ..., \mathcal{C}_l$ so that $\mathcal{C}_i$ starts on the i-th a-edge of $q_1 \cup q_3$, l_1 is the number of a-edges on q_1. Let U_i be the label of the top side of $\mathcal{C}_i$ and V_i be the label of the bottom side of $\mathcal{C}_i$ (see Figure 17). Without loss of generality we can assume that $l_1 > 1$.

Since $\phi(q_1) = \alpha_{\tau^{\pm 1}}(u)$, $\phi(q_3) = \alpha_{\tau'^{\pm 1}}(v)$, $|U_1|$ is bounded by $|p| + 1$, $|V_1|$ is at most $4|U_1|$ (see relations (2.8)), U_2 is bounded by $2 + |V_1| + |p|$, $|V_2| \leq 4|U_2|$, etc. Hence the lengths of all bands $\mathcal{C}_1, ..., \mathcal{C}_{l_1}$ are recursively bounded in terms of $|p|$. Similarly the lengths of all bands $\mathcal{C}_l, \mathcal{C}_{l-1}, ..., \mathcal{C}_{l_1+1}$, which start on q_3, are recursively bounded in terms of $|p|$. Hence indeed, the total number of cells in Δ is recursively bounded in terms of $|p|$.

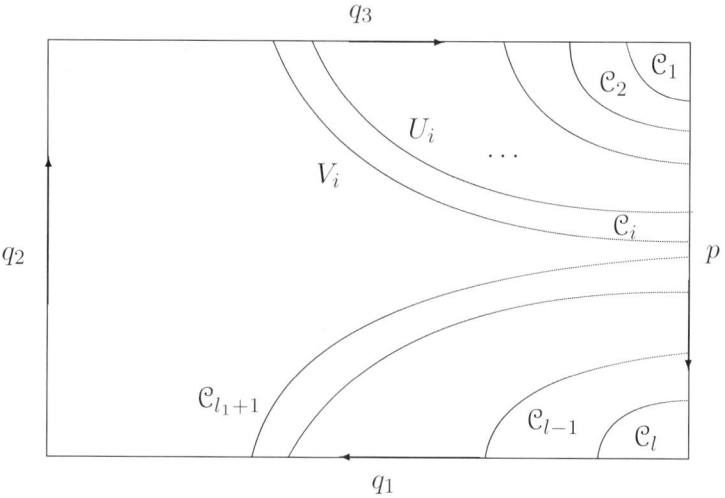

Figure 17.

Since every a-band in Δ ends in p, the number of double edges of the boundary (i.e. edges which appear twice in the contour of the diagram and do not belong to the boundaries of cells of the diagram) cannot exceed a constant times p. Hence the length of the contour of Δ does not exceed a constant times the number of cells in Δ plus a constant times $|p|$. Thus $|q_1 q_2 q_3|$ is recursively bounded in terms of $|p|$. □

5 The word problem in $\mathcal{H}_1$

Recall that $\mathcal{H}_1$ denotes the group given by all generators of $\mathcal{H}$, by all its defining relations, except for the hub Σ, and by all $\mathcal{G}$-relations.

Lemma 5.1. *The word problem in $\mathcal{H}_1$ is decidable.*

Proof. It suffices to find an effective upper bound for the number of cells and perimeters of all $\mathcal{G}$-cells in a reduced van Kampen diagram Δ over $\mathcal{H}_1$, depending on the perimeter $|\partial(\Delta)|$ of Δ. Indeed, since $\mathcal{G}$ is embedded into $\mathcal{H}$ by Lemma 3.9, $\mathcal{G}$ is embedded into $\mathcal{H}_1$ (which satisfies fewer relations than $\mathcal{H}$ but which does satisfy all $\mathcal{G}$-relations. If we know a bound for the perimeters of all $\mathcal{G}$-cells, we can list them all since the word problem in $\mathcal{G}$ is decidable. So in order to check if a word w is equal to 1 in $\mathcal{H}_1$, we would have to consider finitely many van Kampen diagrams with perimeter $|w|$.

First, arguing as in [SBR] and [OlSa2], we bound the number of non-$\mathcal{H}_2$-cells in Δ as follows.

The number of maximal θ-bands of Δ is at most $|\partial(\Delta)|$ since these bands must start and end on the boundary of Δ by Lemma 3.11 (1). Similar upper bounds valid for the number of maximal k-bands, because the diagrams over $\mathcal{H}_1$ have no hubs. Since every (θ, k)-cell is the intersection of a θ-band and a k-band, and two such bands can have at most one intersection by Lemma 3.11 (4), we have a quadratic upper bound (in terms of $|\partial(\Delta)|$) for the number of (θ, k)-cells of Δ.

Since a maximal a-band in Δ cannot be an annulus by Lemma 3.11(3), and cannot connect two $\mathcal{G}$-cells by Lemma 3.12, each maximal a-band in Δ starts or ends either on the boundary of Δ or on the boundary of a (θ, k)-cell. Hence the number of maximal a-bands in Δ is quadratically bounded in terms of $|\partial \Delta|$. Since every (a, θ)-cell is an intersection of a θ-band and an a-band, and two such bands can intersect only once (Lemma 3.11 (5)), the number of (a, θ)-cells in Δ is cubically bounded.

The cubic upper bound is also true for the sum of perimeters of $\mathcal{G}$-cells of Δ. Indeed, each edge on the boundary of a $\mathcal{G}$-cell is the start edge of a a-band which ends either on the boundary of Δ or on the boundary of a (θ, k)-cell.

Notice further, that if Γ is a subdiagram of Δ bounded by a simple loop without θ- and $\mathcal{A}(P_1)$-edges, then Γ is a diagram over $\mathcal{H}_2$. Indeed, Γ has no θ-cells by Lemma 3.11(1). Also Γ has no $\mathcal{G}$-cells, because otherwise the boundary of Γ would contain an $\mathcal{A}(P_1)$-edge (the end edge of a a-band starting on the $\mathcal{G}$-cell), contrary to the assumption.

This implies that every maximal connected subdiagram Γ of Δ without θ- and $\mathcal{A}(P_1)$-edges on the boundary is simply connected. The maximality of Γ implies that every edge of the contour of Γ belongs either to the contour of Δ or to a contour of a non-$\mathcal{H}_2$-cell.

Hence the sum of perimeters of all such Γ does not exceed the sum of perimeters of all non-$\mathcal{H}_2$-cells of Δ plus the perimeter of Δ, i.e. it is effectively bounded. Now we can apply Lemma 4.1, which provides an effective upper bound for the number of $\mathcal{H}_2$-cells in Δ. □

Our main goal is to obtain a similar result for the conjugacy problem in $\mathcal{H}_1$. In this section, we use Lemma 5.1 for getting some preliminary results.

Lemma 5.2. *Let Δ be a reduced diagram over the presentation consisting of relations (2.7), (2.8) and $\mathcal{G}$-relations with contour pq where $\phi(p)$ is a reduced word in $\mathcal{X}$. Then $|p|$ is at most $2|q|2^{\mu_a(q)}$ where $\mu_a(q)$ is the number of a-edges in q, and the number and perimeters of cells in Δ is recursively bounded as function of $|q|$.*

Proof. By Lemmas 3.11 and 3.12, part (2), every maximal a-band in Δ starts and ends on q or starts on q and ends on a $\mathcal{G}$-cell. This implies that the number and the total perimeter of all $\mathcal{G}$-cells in Δ are bounded by $|q|$, since an a-band cannot start and end on $\mathcal{G}$-cells by part (iv) of the definition of reduced diagram and Lemma 3.12(3). Using Lemma 3.12, part (1), we can move the $\mathcal{G}$-cells along the a-bands connecting these cells and q, to q and then remove them from the diagram increasing the length of q by at most a factor of 2. So we can assume that Δ does not contain $\mathcal{G}$-cells, and q is replaced by q_0 with $|q_0| \leq 2|q|$ and the number $n(q_0)$ of a-edges in q_0 for $a \notin \mathcal{A}(P_1)$ does not exceed the number of such edges in q. Thus all maximal a-bands in Δ start and end on q_0.

Let us use induction on the number $n_a(q_0)$ to prove that $|p| \leq |q_0|2^{n(q_0)}$. If $n(q_0) = 0$, then Δ contains no cells and the statement is obvious. Since every cell in Δ belongs to an a-band, there exists a maximal a-band whose side is contained in q_0. Cutting this a-band off Δ, we get a diagram Δ' with boundary pq' where $(q') = n(q_0) - 2$, $|q'| < 4|q_0|$. By the induction hypothesis,

$$|p| \leq |q'|2^{n(q')} < 4|q_0|2^{n(q_0)-2} = |q_0|2^{n(q_0)}.$$

Since we have recursively bounded the number of a-bands in Δ and the length of each a-band, the number of cells in Δ is recursively bounded. □

Lemma 5.3. *Let Δ be a reduced diagram over the presentation consisting of relations (2.7), (2.8) and $\mathcal{G}$-relations. Suppose $\partial(\Delta) = pq$ where $\phi(p)$ is the label of a side of a reduced k-band (the k-band is not contained in Δ, of course). Suppose also that all θ-bands starting on p end on q. Then the number and the perimeters of cells in Δ are recursively bounded in terms of $|q|$.*

Proof. Let $\mathfrak{T}_1, ..., \mathfrak{T}_n$ be the maximal θ-bands starting on p (counted from the beginning of p to the end of p). Then $n \leq |q|$. Let Δ_0 be the subdiagram bounded by $\partial(\Delta)$ and **bot**($\mathfrak{T}_1$) (the one that does not contain $\mathfrak{T}_1$), for every $i = 1, ..., n-1$ let Δ_i be the subdiagram bounded by $\partial(\Delta)$, **top**($\mathfrak{T}_i$) and **bot**($\mathfrak{T}_{i+1}$). Finally let Δ_n be the subdiagram bounded by **top**($\mathfrak{T}_n$) and $\partial(\Delta)$, not containing $\mathfrak{T}_n$. Then for every $i = 0, ..., n$, the a-edges on $\partial(\Delta_i) \cap p$ belong to the initial or terminal segments of $\partial(\Delta_i) \cap p$ of constant length (not exceeding the length of a relator of the form (2.6)). Hence by Lemma 3.11, all but a constant number of maximal a-bands in Δ_i ($i = 0, ..., n$) starting on the boundary of $\mathcal{T}_{i+1}$ ($i = 0, ..., n-1$), end either on q or on a side of $\mathcal{T}_i$ or on the boundary of a $\mathcal{G}$-cell.

Suppose that one of the a-bands in Δ_i starting on a side of $\mathcal{T}_{i+1}$ ($i = 0, 1, ...$) ends on the boundary of a $\mathcal{G}$-cell. Then the label of an a-edge of $\mathcal{T}_{i+1}$ belongs to $\mathcal{A}(P_1)$. This implies that the Θ-edges of $\mathcal{T}_{i+1}$ belong to $\Theta(P_1)$. This means that p is a side of a P_1- or R_1-band (see relations (2.6). Since there are no P_1- or R_1-relations of the form (2.9), the length of p is at most a constant times the number of θ-edges in p, i.e. does not exceed a constant times $|q|$. Thus the perimeter of Δ is linearly bounded in terms of $|q|$, so the statement of the lemma follows from Lemma 5.1.

Thus we can assume that none of the a-bands in Δ_i starting on a side of $\mathcal{T}_{i+1}$ ends on the boundary of a $\mathcal{G}$-cell. Therefore all but a constant number of the a-bands in Δ_i starting on the boundary of $\mathcal{T}_{i+1}$ end on $\mathcal{T}_i$ or on q. Therefore the length of $\mathcal{T}_1$ is recursively bounded in terms of $|q|$, and for each of each $i = 2, ..., n$, $|\mathcal{T}_i|$ is recursively bounded in terms of $|q|$ and $|\mathcal{T}_{i-1}|$. Hence the lengths of all $\mathcal{T}_i$ are recursively bounded in terms of $|q|$.

Thus for every $i = 0, ..., n$, the diagram Δ_i consists of cells corresponding to relations (2.8), and $\mathcal{G}$-relations the word $\phi(\partial(\Delta_i) \cap p)$ has the form uvu' where u, u' have lengths bounded by a constant, and v is a word in $\mathcal{X}$, and the complement of $\partial(\Delta_i) \cap p$ to $\partial(\Delta_i)$ has length recursively bounded in terms of $|q|$. Therefore by Lemma 5.2, the length of $\partial(\Delta_i) \cap p$ is also recursively bounded in terms of $|q|$, and the number and perimeters of cells in Δ_i are also recursively bounded. This implies that $|p|$ and the number of cells in Δ are recursively bounded in terms of $|q|$. □

Let us introduce the following desirable properties of a boundary component p (inner or outer contour) of an annular diagram Δ over $\mathcal{H}_1$:

(R1) p does not contain the boundary of a $\mathcal{G}$-cell as a subpath;

(R2) no a-band in Δ starts and ends on p;

(R3) no θ-band in Δ starts and ends on p;

(R4) no k-band starts and ends on p.

The next lemma shows that in some cases we can achieve these properties without making too many changes in the diagram.

For every reduced path p in a diagram over $\mathcal{H}_1$, denote by $\mu_k(p)$ (resp. $\mu_\theta(p), \mu_a(p), \mu_x(p)$) the number of all k-edges (resp. θ-edges, a-edges, x-edges) in p. The vector $(\mu_k(p), \mu_\theta(p), \mu_a(p), \mu_x(p))$ will be called the *type* of p. We order types lexicographically.

Lemma 5.4. *Let Δ be a reduced annular diagram over $\mathcal{H}_1$ with boundary components p and q. Then by removing some cells from Δ, one can construct an annular diagram Δ_1 with boundary components p_1 and q satisfies* condition (R4), such that $|p_1|$, the number and perimeters of cells in the diagram $\Delta \backslash \Delta_1$ are recursively bounded in terms of $|p|$.

Proof. Indeed, suppose that Δ contains a k-band starting and ending, say, on p. Consider such a k-band $\mathcal{B}$ which is the closest to p, so that the subdiagram Γ of Δ bounded by **top**$(\mathcal{B})$ and p contains no k-bands. By Lemma 3.11, no θ-band can start and end on $\mathcal{B}$. Therefore, by Lemma 5.3, the number and perimeters of cells in Γ are recursively bounded in terms of $|p|$. Removing Γ and $\mathcal{B}$ we obtain a diagram with smaller number of k-edges on the boundary. Thus we can continue that operation at most $|p|$ times and obtain the desired diagram Δ_1. □

Lemma 5.5. *Let Δ be a reduced annular diagram over $\mathcal{H}_1$ with a boundary components p and q, and p satisfy the condition (R4). Then there exists another (annular) diagram Δ' with contours p' and q satisfying (R1), (R2), (R3), (R4) and such that:*

(R0) The path p' satisfies the following properties:

- *The type of p' is not greater than the type of p, and $|p'|$ is recursively bounded in terms of $|p|$.*

- *The word $\phi(p')_\theta$ is freely conjugate of the word $\phi(p)_\theta$;*

- *If p does not contain k-edges then the word $\phi(p')_a$ is a conjugate of the word $\phi(p)_a$ modulo the $\mathcal{G}$-relations from Δ;*

- *The words $\phi(p)$ and $\phi(p')$ are conjugate modulo relations used in Δ. The number and lengths of relations used to deduce these conjugacies are recursively bounded in terms of $|p|$;*

Proof. Let us describe three operations which can be applied to Δ.

1. If a p contains a subpath which is the boundary of a $\mathcal{G}$-cell, we can remove that cell from Δ lowering the length of the boundary. This operation does not increase the number of k-edges, θ-edges or a-edges on this boundary component of the diagram. The θ-projections of the labels of the boundary components of the diagram do not change. Notice that the resulting diagram satisfies (R0).

2. Suppose that an a-band $\mathcal{C}$ in Δ starts and ends on p. Let Δ_1 be the subdiagram of Δ bounded by $\mathcal{C}$ and p. It has no k-edges by property (R4) and Lemma 3.11, because $\mathcal{C}$ cannot possess a k-cell. Since the top/bottom of $\mathcal{C}$ has no a-edges, we can apply lemma 3.12, as in the proof of Lemma 5.2, and remove all $\mathcal{G}$-cells of Δ_1 increasing the length of the p-part of its boundary at most twice, but not changing $\mathcal{C}$.

Every maximal a-band in Δ_1 starts and ends on p since Δ_1 has neither k- nor $\mathcal{G}$-cells. Therefore we could find an a-band $\mathcal{C}$ that starts and ends on, say, p, and such that there are no cells between the top path of the band

and p. Then the length of this band would be bounded by $|p|$. Notice that the number of θ-edges on the top path of an a-band is equal to the number of θ-edges on the bottom path of the band, and the corresponding numbers of $\mathcal{X}$-edges can differ at most 4 times. Therefore we could delete that band and reduce the number of a-edges in $\partial(\Delta)$ but not increase the number of θ-edges there and preserve property (R0) (in particular, since $\mathcal{C}$ connects two consecutive a-edges on p, $\phi(p')_a$ is obtained from $\phi(p)_a$ by removing a subword of the form aa^{-1}, so $\phi(p)_a = \phi(p')_a$ in the free group).

3. Suppose that a θ-band $\mathcal{T}$ in Δ starts and ends on p, and there are no θ-cells in the simply connected subdiagram Γ of Δ between the top side of that band an p. If $\mathbf{top}(\mathcal{T})$ is a part of p, then we can remove $\mathcal{T}$ as in part 2, reducing the number of θ-edges on the boundary of the diagram, cancelling two consecutive mutually inverse θ-letters in the cyclic word $\phi(p)_\theta$, and not increasing the number of k-edges there, thus preserving (R0) (notice, in particular, that if there are no k-edges on p, then $\phi(\mathbf{top}(\mathcal{T}))_a \equiv \phi(\mathbf{bot}(\mathcal{T}))_a$, whence $\phi(p')_a \equiv \phi(p)_a$).

Assume now that there is a $\mathcal{G}$-cell Π in Γ, connected with $\mathcal{T}$ by an a-band. This a-band has no cells, since Γ contains no θ-cells. Hence the contour $p_\Pi q_\Pi$ of Π has a common part q_Π of positive length with $\mathcal{T}$. Then , by relations (2.7), $\phi(p_\Gamma)$ and $\phi(q_\Gamma)$ commute with the $\theta(P_1)$-letters labelling the θ-edges of the subband $\mathcal{T}_0$ of $\mathcal{T}$ with top side q_Π. Therefore one can construct an auxiliary θ-band $\mathcal{T}'$, whose top label is obtained from that of $\mathcal{T}$ by replacing the subwords $\phi(q_\Pi)$ by $\phi(p_\Pi)$. Then one can paste $\mathcal{T}'$ to $\Gamma\backslash(\Pi\cup\mathcal{T}_0)$ reducing the number of $\mathcal{G}$-cells in Γ, but preserving property (R0). Notice that this transformation does not affect the boundary labels of Δ.

Hence one may assume further that no $\mathcal{G}$-cell of the reduced subdiagram Γ is connected with $\mathcal{T}$ by an a-band. Therefore the sum of perimeters of all $\mathcal{G}$-cells in Γ is not greater than the length of the p-part p_Γ of its boundary. Then, as in part 2, one can remove all $\mathcal{G}$-cells from Γ increasing the number of a-edges in p_Γ at most twice and preserving (R0). Thus we may assume that Γ has no $\mathcal{G}$-cells.

Let $\mathcal{B}_1, ..., \mathcal{B}_r$ be all the maximal a- and k-bands in Γ starting on p, of non-zero length. Each of the bands $\mathcal{B}_1, \ldots \mathcal{B}_r$ ends either on $\mathcal{T}$ or on p (an a-band $\mathcal{B}_i$ cannot end on a k-cell not from $\mathcal{T}$ because every k-cell containing an a-edge also contains a θ-edge). Therefore there is one of them (if $r > 0$), say $\mathcal{B}_1$, such that the top or bottom side p^1 of $\mathcal{B}_1$ has no edges in common with the sides of $\mathcal{B}_2, \ldots, \mathcal{B}_r$ and has at most 4 common x-edges with $\mathcal{T}$. ($\mathcal{T}$ can have at most 2 consecutive x-letter in its boundary label, as it follows from the form of θ-relations.) Since all the edges of p^1 (with at most 4 exceptions) belong to p, the length of $\mathcal{B}_1$ is linearly bounded. Removing this band from Γ does not violate property (R0). Considering the diagrams $\Gamma\backslash\mathcal{B}_1, \Gamma\backslash(\mathcal{B}_1\cup\mathcal{B}_2)\ldots$ we can recursively bound the length of every $\mathcal{B}_2, \ldots \mathcal{B}_r$, because the number r was linearly bounded above, and remove them all from Γ. But the case where Γ has no cells was considered in the beginning.

Now repeating operations 1, 2, 3 recursively bounded (in terms of $|p|$) number of times we will get the desired annular diagram. □

We call a word W *θ-minimal* if the number of θ-letters in W are not greater than the similar numbers for any word W' which is a conjugate of W in $\mathcal{H}_1$. Similarly we define *k-minimal* words. A word W is called *θk-minimal* if it is both θ- and k-minimal. A word W is said to be *θa-minimal*, if the number of θ-letters and the number of non-$\mathcal{A}(P_1)$ a-letters in W are not greater than the corresponding numbers for any word W' which is a conjugate of W modulo the relations of $\mathcal{H}_1$ except for (θ, k)-relations. A boundary component of an (annular) diagram Δ over $\mathcal{H}_1$ is called minimal if its label is θk-minimal and, if Δ has no (θ, k)-cells, it is also θa-minimal. Let us consider reduced annular diagrams with boundary labels W and W' for various words W'. The immediate analysis of the transformations used in Lemmas 5.4 and 5.5 proves

Lemma 5.6. *For any word W representing an element of $\mathcal{H}_1$, there is an annular diagram Δ_0 over $\mathcal{H}_1$ with boundary labels W and W_0, where W_0 is θk-minimal (θa-minimal, and Δ_0 has no (θ, k)-cells), such that the length of W_0, the number, and the perimeters of cells in Δ_0 are recursively bounded as functions of $|W|$.*

6 Some special diagrams

6.1 θ-bands and trapezia

Let W be a reduced word in the generators of $\mathcal{H}$. The *base* of W is the projection of W onto $\tilde{\mathcal{K}}$.

Let $\tau \in \mathcal{S} \cup \bar{\mathcal{S}}$, and $W \equiv U_1 z_1 U_2 \ldots z_n U_{n+1}$ a word where $z_1, \ldots, z_n$ are basic letters, $U_1, \ldots, U_{n+1}$ are words in $\mathcal{A} \cup \bar{\mathcal{A}} \cup \mathcal{X}$. Denote $z_0 = (z_1)_-$, $z_{n+1} = (z_n)_+$. For every word V let $V_{\setminus x}$ be the word V with all x-letters deleted. We say that the word W is *τ-regular* if

- for every $i = 1, \ldots, n+1$ the word $z_{i-1}(U_i)_{\setminus x} z_i$ is admissible for $\mathcal{S} \cup \bar{\mathcal{S}}$ and in the domain of τ;

- if $\tau \in \mathcal{S}$, then $W \equiv \alpha_\tau(W_{\setminus x})$, and if $\tau \in \bar{\mathcal{S}}$, then W does not contain x-letters.

Thus for every admissible word W of $\mathcal{S}$ to which $\tau \in \mathcal{S}$ is applicable the word $\alpha_\tau(W)$ is τ-regular, and if $\tau \in \bar{\mathcal{S}}$ is applicable to an admissible word W of $\bar{\mathcal{S}}$ then W itself is τ-regular.

The following lemma immediately follows from the form of θ-relations from $\mathcal{R}'$.

Lemma 6.1. *Let $\mathcal{B}$ be a reduced $\Theta(\tau)$-band for some $\tau \in \mathcal{S} \cup \bar{\mathcal{S}}$. Then the reduced labels of the top and the bottom paths of $\mathcal{B}$ have the same bases which will be called* the base *of $\mathcal{B}$, and are τ- or τ^{-1}-regular words.*

For future references it is convenient to list all possible 2-letter bases of τ-regular words (i.e. 2-letter subwords of $\tilde{\Sigma}^{\pm 1}$).

Lemma 6.2. *Let yz be a 2-letter base of a reduced $\Theta(\tau)$-band, $y, z \in \tilde{\mathcal{K}} \cup \tilde{\mathcal{K}}^{-1}$. Then yz or $(yz)^{-1}$ has one of the following forms: $\overleftarrow{L_j}L_j$, L_jP_j, P_jR_j, $R_j\overrightarrow{R_j}$, $z_jz_j^{-1}$, $z_j^{-1}z_j$ where $z \in \{K, L, P, R\}$, $j = 1, ..., N$. In addition if $\tau \in \bar{\mathcal{S}}$, yz cannot have the form $z_1 z_1^{-1}$ or $z_1^{-1}z_1$ where $z \in \{L, R, P\}$, and it cannot be equal to either $K_1 K_1^{-1}$ or to $K_2^{-1}K_2$. Moreover if τ locks zz_+-sectors then yz cannot have the form zz^{-1}.*

Proof. Only the "moreover" statement needs explanation. If τ locks zz_+-sectors, then there are no corresponding relations of the form (2.7). Hence if the base has the form zz^{-1}, the band has two consecutive mirror image (θ, z)-cells that cancel. This contradicts the assumption that $\mathcal{B}$ is reduced. □

The following lemma immediately follows from relations (2.6), (2.7).

Lemma 6.3. *Let yz be the base of a $\Theta(\tau)$-band $\mathcal{B}$. Let V be the label of the top path of $\mathcal{B}$, V' be the label of the bottom path of $\mathcal{B}$. Let $V = \alpha_{\tau \pm 1}(W_1 y(r, i) W_2 z(r, i) W_3)$, $V' = \alpha_{\tau \mp 1}(W_1' y(r', i') W_2' z(r', i') W_3')$ provided $\tau \in \mathcal{S}$ and let $V = W_1 y(r, i) W_2 z(r, i) W_3$, $V' = W_1' y(r', i') W_2' z(r', i') W_3'$ provided $\tau \in \bar{\mathcal{S}}$. Let $u(y_-), v(y), u(y), v(z)$ be the words associated with τ and defined in Section 2.4. Then, in the free group, $W_1' = W_1 v(y_-)$, $W_2' = u(y)W_2 v(y)$, $W_3' = u(z)W_3$.*

A *trapezium* (see Figure 18) is a reduced van Kampen diagram Δ over the group $\mathcal{H}_1$ whose boundary path is factorized as $p_1 p_2 p_3 p_4$, where

(1) the paths p_2 and p_4 have no θ-edges;

(2) p_1 and p_3^{-1} are sides of k-bands $\mathcal{B}$ and $\mathcal{B}'$ starting on p_4 and ending on p_2 (i.e. $p_1 = \mathbf{top}(\mathcal{B})$, $p_3^{-1} = \mathbf{bot}(\mathcal{B}')$).

(3) every maximal k-band in Δ contain at least one θ-cell and connects p_2 and p_4.

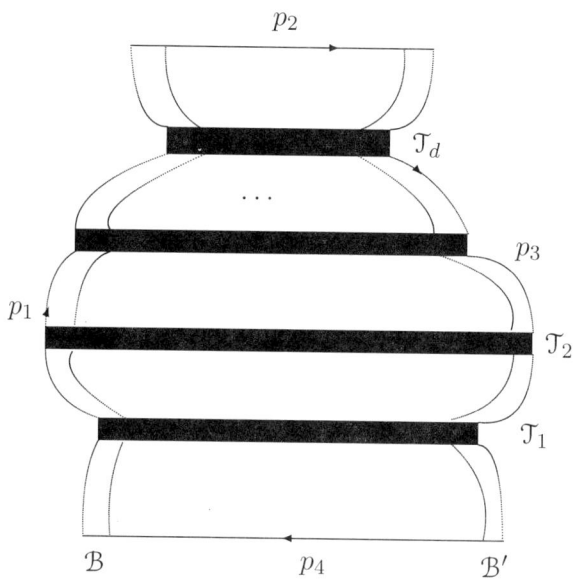

Figure 18.

By definition, p_2 and p_4 are the *top* and the *bottom* of the trapezium, respectively. As follows from Lemma 3.11, every maximal θ-band of the trapezium connects p_1 and p_3. We shall usually enumerate maximal θ-bands starting on p_1 from the bottom to the top: $\mathcal{T}_1, ..., \mathcal{T}_d$. The number of them is called the *height* of the trapezium. The paths p_1, p_2, p_3, p_4, the bands $\mathcal{B}$, $\mathcal{B}'$ and $\mathcal{T}_1, ..., \mathcal{T}_d$ form the *data* associated with the trapezium. By Lemma 3.11, each of these k-bands intersects each of the θ-bands exactly once. Therefore by Lemma 6.1 the bases of the θ-bands $\mathcal{T}_1, ..., \mathcal{T}_d$ are the same. Hence the base of the label of the bottom p_4 will be called the *base* of the trapezium (it is equal to the base of each of $\mathcal{T}_i$). Notice that every k-band is a trapezium with a 1-letter base.

A reduced annular diagram Δ is called a *ring* with boundary components p_2 and p_4 if:

(1) the labels of p_2 and p_4 contain no θ-edges;
(2) the boundaries p_2 and p_4 are minimal;
(3) Δ has at least one (k, θ)-cell.

It follows from the definition, that a ring Δ contains at least one k-band having a (θ, k)-cell, every maximal k-band of Δ connects p_2 and p_4, and a ring is obtained from a trapezium by identifying two bands $\mathcal{B}$ and $\mathcal{B}'$ (in this case $\mathcal{B}$ must be a copy of $\mathcal{B}'$ in the definition of trapezium). The θ-bands in a ring are annuli surrounding the hole of the ring. The top and bottom paths of the trapezium turn into the *outer* and *inner* paths of the ring. The base and the height of a ring are defined as for trapezia, but the base is considered as a cyclic word.

If the bands $\mathcal{B}$ and $\mathcal{B}'$ from the definition of trapezia are allowed to be a-bands, $a \notin \mathcal{A}(P_1)$, one obtains the definition of *quasitrapezium*.

A *quasiring* is either a ring or a reduced annular diagram with contours p and q such that (1) Δ contains no (θ, k)-cells but contains a (θ, a)-cell with $a \neq \mathcal{A}(P_1)$; (2) the labels of the contour p and q have neither θ nor k-letters; (3) the boundaries p and q are minimal.

Let $\mathcal{T}_i$, $i = 1, ..., d$ be a $\Theta(\tau_i)$-band where $\tau_i \in \mathcal{S} \cup \bar{\mathcal{S}}$. Then the word $h = \tau_1 \tau_2 ... \tau_d$ is called the *history* of the trapezium (ring). Notice that the length of the history is equal to the height of the trapezium (ring). By Lemma 3.11, the history of any k-band in a trapezium (ring) is equal to the history of the trapezium (ring).

The definitions of the base and the history of a quasitrapezium (quasiring) are similar to those for trapezia (ring), but the base of a quasitrapezium can be empty. Of course if the base of a quasiring is not empty then it is a ring.

Lemma 6.4. *For any reduced annular diagram Δ over $\mathcal{H}_1$ whose contours p and q do not contain θ-edges, there exists a ring or a quasiring, or a diagram with minimal boundaries and having no cells, Δ' whose contours p' and q' satisfy properties (R0)-(R4) of Lemma 5.5, in which the lengths p' and q' are recursively bounded in terms $|p| + |q|$, and there exists a conjugacy diagram for $\phi(p)$ (for $\phi(q)$) and $\phi(p')$ (and $\phi(q')$) with recursively bounded number and perimeters of cells.*

Proof. By Lemma 5.6 one can assume that p is a minimal boundary. We may also assume that q enjoys the same property.

If we have a (θ, k)-cells in Δ, then the diagram is a roll. So we assume that there are no (θ, k)-cells in it.

If there is a k-edge in $\partial \Delta$, then a maximal k-band, containing no θ-cells, connects p and q. Therefore Δ has no θ-edges by Lemma 3.11. Hence it is a diagram over the free product of the group $\mathcal{H}_2$ and $\mathcal{G}$, and the statement follows from Lemma 4.4.

A reduced diagram cannot possess a θ-annulus consisting of $(a(P_1), \theta)$-cells (with equal boundary labels). Hence, if p and q have neither θ- nor k-edges, then Δ is a *quasiring* as desired, or it has no θ-cells. If Δ has no θ-cells, then we can repeat the argument of the previous paragraph. $\square$

Lemma 6.5. *(1) Let $y_1 ... y_s$ be the base of a trapezium Δ whose top and bottom labels have no letters from $\mathcal{A}(P_1)$. Assume the words $y_1 ... y_{s-1}$, $(y_2 ... y_s)^{-1}$ contain no occurrence of the positive letter P_1, and the words $y_2 ... y_s$, $(y_1 ... y_{s-1})^{-1}$ contain no occurrence of the positive letter R_1. Then Δ has no $\mathcal{G}$-cells.*

(2) If the base of a quasiring does not contain $P_1^{\pm 1}$ and $R_1^{\pm 1}$ and the inner and outer paths of the ring do not contain $\mathcal{A}(P_1)$-edges then the ring does not have $\mathcal{G}$-cells.

Proof. (1) Indeed, by Lemma 3.12, we can assume that every $\mathcal{A}(P_1)$-band starting on a $\mathcal{G}$-cell must end either on the boundary of the diagram or on the boundary of a P_1-cell or on the boundary of a R_1-cell. The first case would imply that the bottom or the top of the trapezium contains $\mathcal{A}(P_1)$-edges. The second case would imply that one of the letters $y_1, ..., y_{s-1}, y_2^{-1}, ..., y_s^{-1}$ is P_1. The third case would imply that one of the letters $y_2, ..., y_s, y_1^{-1}, ..., y_{s-1}^{-1}$ is R_1.

Statement (2) is proved similarly. $\square$

6.2 Trapezia with 2-letter bases

Let Δ be a trapezium with a 2-letter base yz. We use the notations from Section 6.1 and from the definition of a trapezium for the data of Δ. Let $V_0 \equiv \phi(p_4^{-1})$, $U_{d+1} \equiv \phi(p_2)$, and for all $i = 1, ..., d$ let $U_i \equiv \phi(\mathbf{bot}(\mathcal{T}_i))$, $V_i \equiv \phi(\mathbf{top}(\mathcal{T}_i))$.

For every word W, we denote the projection $W_{\mathcal{A} \cup \bar{\mathcal{A}}}$ of W onto the alphabet $\mathcal{A} \cup \bar{\mathcal{A}}$ by W_a. Similarly we denote $W_{\Theta \cup \bar{\Theta}}$ by W_θ.

Lemma 6.6. *(1) Suppose that Δ does not have $\mathcal{G}$-cells. Then $(U_1)_a = (V_0)_a$, $(V_d)_a = (U_{d+1})_a$ in the free group, and $(V_i)_a \equiv (U_{i+1})_a$, $i = 1, ..., d-1$.*

(2) If Δ contains $\mathcal{G}$-cells then $(U_1)_a = (V_0)_a, (V_d)_a = (U_{d+1})_a$, $(V_i)_a = (U_{i+1})_a$, $i = 1, ..., d-1$, modulo the $\mathcal{G}$-relations.

Proof. Let Γ_0 be the subdiagram of Δ bounded by $p_1, p_3, p_4, \mathbf{bot}(\mathcal{T}_1)$, Γ_d be the subdiagram bounded by $p_1, p_3, p_2, \mathbf{top}(\mathcal{T}_d)$, and for every $i = 1, ..., d-1$ let Γ_i be the subdiagram bounded by $p_1, p_3, \mathbf{top}(\mathcal{T}_i)$ and $\mathbf{bot}(\mathcal{T}_{i+1})$.

(1) Suppose that Δ does not contain $\mathcal{G}$-cells. A maximal a-band in Γ_i which starts on a side of a θ-band cannot end up on the same side because otherwise the top/bottom labels of the θ-band would not be reduced. This implies that $(V_1)_a \equiv (U_2)_a, ..., (V_{d-1})_a \equiv (U_d)_a$.

Also every a-band which starts on $\mathbf{bot}(\mathcal{T}_1)$ ends on p_4, every a-band which starts on $\mathbf{top}(\mathcal{T}_d)$ ends on p_2. If an a-band starts and ends on p_4 (resp. p_2) then the label of the a-path between the start and end edges of that band must be freely equal to 1. Hence $(V_0)_a = (U_1)_a$ and $(V_d)_a = (U_{d+1})_a$ in the free group.

(2) Cells in Γ_i do not contain θ-edges. Hence they correspond to either relations of the form (2.8) or $\mathcal{G}$-relations. If we remove all $\mathcal{X}$-letters in relations (2.8), the relations become trivial. Thus if we collapse all $\mathcal{X}$-edges in Γ_i, $i = 0, ..., d$, that diagram becomes a diagram with $\mathcal{G}$-cells only. (More precisely, to the boundary label of Γ_i, we apply the homomorphism which preserves letters $a_i(P_1)$, $1 \leq i \leq m$, and sends other a-letters and $\mathcal{X}$-letters to 1.) Since the portions of p_1 and p_3 on the contour of Γ_i consist of $\mathcal{X}$-edges, these portions collapse to vertices. Now the van Kampen lemma implies that $(U_1)_a = (V_0)_a, (V_d)_a = (U_{d+1})_a$, $(V_i)_a = (U_{i+1})_a$, $i = 1, ..., d-1$, modulo the $\mathcal{G}$-relations, as required. $\square$

The following lemma immediately follows from the fact that there are no $(P_j, \mathcal{X})$- and $(R_j, \mathcal{X})$-relations of the form (2.9).

Lemma 6.7. *One of the sides of every P_j-band (R_j-band) in a van Kampen diagram over $\mathcal{H}_1$ contains a- and θ-edges only.*

Lemma 6.8. *(1) If the base of a trapezium Δ contains a P_j- or a R_j-letter, then the history is a reduced word.*

(2) The history of an arbitrary ring is a reduced word.

Proof. (1) Let $\tau\tau^{-1}$ be a 2-letter subword of the history of a trapezium Δ. Consider the corresponding pair $\mathcal{T}_i$, $\mathcal{T}_{i+1}$ of consecutive θ-bands in Δ. The intersections of these bands with a P_j- or R_j-band $\mathcal{C}$ of Δ are two cells π, π' which are consecutive cells in $\mathcal{C}$ by Lemma 6.7. Since π and π' correspond to τ and τ^{-1}, they form a reducible pair of cells which contradicts the assumption that Δ is reduced.

(2) Let Δ be a ring. Without loss of generality we can assume that the history of Δ is $\tau\tau^{-1}$, so Δ is of height 2. By (1), we can assume that the base of Δ has neither P_j- nor R_j-letters. We can also suppose that Δ is constructed of two θ-bands $\mathcal{T}_1$, $\mathcal{T}_2$ and a of the subdiagram Γ bounded by $\mathbf{top}(\mathcal{T}_1)$ and $\mathbf{bot}(\mathcal{T}_2)$, which have no θ-cells.

By Lemma 6.5, applied to all subtrapezia of Δ with 2-letter bases, Δ has no $\mathcal{G}$-cells. Therefore, by Lemma 6.6 (1) we have that the words obtained from the labels of $\mathbf{top}(\mathcal{T}_1)$ and $\mathbf{bot}(\mathcal{T}_2)$ after deletion of all $\mathcal{X}$-letters, are freely equal. The same is true for the labels of $\mathbf{bot}(\mathcal{T}_1)$ and $\mathbf{top}(\mathcal{T}_2)$ by Lemma 6.3, since bands $\mathcal{T}_1$ and $\mathcal{T}_2$ correspond to mutually inverse rules. Hence $\phi(\mathbf{bot}(\mathcal{T}_1)) \equiv \phi(\mathbf{top}(\mathcal{T}_2))$ by Lemma 6.1. But this contradicts the assumption that Δ is reduced (see part (iv) of the definition of non-reduced diagram). □

6.3 Trapezia simulate the work of S-machines

The following lemma shows that every trapezium simulates the work of $\mathcal{S} \cup \bar{\mathcal{S}}$.

Lemma 6.9. *Let Δ be a trapezium of height d with θ-bands $\mathcal{T}_1, ..., \mathcal{T}_d$, W (resp. W') be the projection of the label of the bottom (resp. top) path of Δ onto $\mathcal{A} \cup \bar{\mathcal{A}} \cup \mathcal{K} \cup \bar{\mathcal{K}}$, U_i (resp. V_i), $i = 1, ..., d$, be the projection of $\phi(\mathbf{bot}(\mathcal{T}_i))$ (resp. $\phi(\mathbf{top}(\mathcal{T}_i))$) on the same set. Let $h = \tau_1 \cdot ... \cdot \tau_d$ be the history of Δ. Then U_i, V_i, $i = 1, ..., d$, are admissible words for $\mathcal{S} \cup \bar{\mathcal{S}}$. In addition*

$$W = U_1 (mod\ \mathcal{G}), V_1 = U_1 \circ \tau_1, U_2 = V_1 (mod\ \mathcal{G}), ...,$$
$$V_d = U_d \circ \tau_d, W' = V_d (mod\ \mathcal{G}), \quad (6.1)$$

(here mod $\mathcal{G}$ means that the equality is true modulo $\mathcal{G}$-relations, other equalities are true in the free group). If Δ does not contain $\mathcal{G}$-cells then one can remove (mod $\mathcal{G}$) from the previous statements. We also have $||U_i| - |V_i|| \leq Bc$

where B is the length of the base of Δ and c is the maximum of lengths of words in $\bar{\mathcal{E}}$.

Proof. Suppose first that Δ contains just one θ-band $\mathcal{T}$. Let U and V be the projections of $\phi(\mathbf{bot}(\mathcal{T}))$ and $\phi(\mathbf{top}(\mathcal{T}))$ respectively on $\mathcal{A} \cup \bar{\mathcal{A}} \cup \mathcal{K} \cup \bar{\mathcal{K}}$. Lemmas 6.1 and 6.3 imply that U and V are admissible words for $\mathcal{S} \cup \bar{\mathcal{S}}$ and $V = U \circ \tau$. By Lemma 6.6 $W = U$ and $W' = V$ in the free group if Δ does not have $\mathcal{G}$-cells, otherwise these equalities hold modulo $\mathcal{G}$-relations. The fact that $||V_i| - |U_i||$ follows from Lemma 6.3. Now the proof can be finished by a simple induction on the height of Δ. $\square$

We shall call (6.1) the *computation associated with trapezium* Δ. The *length* of the computation is the height of the trapezium.

Lemma 6.10. *(1) Let x_1, y_1, x_2, y_2 be letters in $\mathcal{X}$, $x_1 \neq y_1^{\pm 1}$, $x_2 \neq y_2^{\pm 1}$, and let U_1, U_2 be non-empty reduced words which are products of fourth powers of letters from $\mathcal{X}$. Then $U_1 x_1 y_1^{-1} U_2 y_2 x_2^{-1} \neq 1$ in the free group.*

(2) Let $x_1, x_2 \in \mathcal{X}^{\pm 1}$, let U and $x_1 U x_2$ be products of fourth powers of letters from $\mathcal{X}$. Then $x_2 \equiv x_1^{-1}$ and U is a power of x_1.

(3) Let $x_1, x_2 \in \mathcal{X}^{\pm 1}$, let U be a non-empty power of a letter $x \in \mathcal{X}$, $x_1 U x_2$ is a product of fourth powers of letters of $\mathcal{X}$. Then $x_1, x_2 \in \{x, x^{-1}\}$.

Proof. (1) Suppose $U_1 x_2 y_2^{-1} U_2 y_1 x_1^{-1} = 1$ in the free group. Then $U_1 x_2 y_2^{-1} = x_1 y_1^{-1} U_2^{-1}$. Since U_1 is a nonempty product of fourth powers of letters, the reduced form of $U_1 x_2 y_2^{-1}$ ends with $x_2^{\pm 1} y_2^{-1}$. But since U_2 is a non-empty product of fourth powers of letters, the word $x_1 y_1^{-1} U_2$ ends with a third power of a letter, a contradiction.

The other statements of the lemma are proved similarly. $\square$

A (quasi-)trapezium or a ring Δ is of the *first (second, mixed) type* if its history is a word in $\mathcal{S}$ (resp. $\bar{\mathcal{S}}$, $\mathcal{S} \cup \bar{\mathcal{S}}$ but not $\mathcal{S}$ or $\bar{\mathcal{S}}$).

Lemma 6.11. *Let Δ be a quasitrapezium of the first or mixed type with history of length 2. Let $\mathcal{T}_1, \mathcal{T}_2$ be the two maximal θ-bands of Δ counting from the bottom up. Assume that $\mathcal{T}_1$ is a Θ-band. Let V be the label of $\mathbf{top}(\mathcal{T}_1)$. Then V_a cannot contain subwords $a_i(z)^{-1} a_{i'}(z)$ if $z \in \{K_j, L_j\}$ and it cannot contain subwords $a_i(z) a_{i'}(z)^{-1}$ if $z = R_j$.*

Proof. We shall consider only the case $z = L_j$ because other cases are similar. Suppose that V_a contains a subword $a_i(L_j)^{-1} a_{i'}(L_j)$. Let $\mathcal{B}_1, \mathcal{B}_2$ be the two neighbor a-bands starting on $\mathbf{top}(\mathcal{T}_1)$ and ending on $\mathbf{bot}(\mathcal{T}_2)$ corresponding to this subword, i.e. $\mathcal{B}_1$ is a $a_i(L_j)^{-1}$-band, $\mathcal{B}_2$ is a $a_{i'}(L_j)$-band.

Let Γ be the subdiagram of Δ which is situated between $\mathcal{B}_1$ and $\mathcal{B}_2$, i.e. it is bounded by $\mathbf{bot}(\mathcal{B}_1)$ and $\mathbf{top}(\mathcal{B}_2)$, a portion of $\mathbf{top}(\mathcal{T}_1)$, and a portion of $\mathbf{bot}(\mathcal{T}_2)$. Let $\partial(\Gamma) = u_1 q u_2 p^{-1}$ be the decomposition of the boundary of Γ where $u_1 = \mathbf{bot}(\mathcal{B}_1)$, $u_2 = \mathbf{top}(\mathcal{B}_2)^{-1}$.

Notice that the start (respectively end) edges of $\mathcal{B}_1$ and $\mathcal{B}_2$ on $\mathbf{top}(\mathcal{T}_1)$ (respectively $\mathbf{bot}(\mathcal{T}_2)$) belong to two different cells π_1, π_1' (resp. π_2, π_2') because in every θ-cell, edges labelled by a-letters in opposite exponents are always separated by a θ-edge: this is obvious for relations (2.7) and relations (2.6) corresponding to rules not from $\mathcal{S}(34) \cup \bar{\mathcal{S}}(34)$; for the remaining relations it follows from the assumption that all words in $\bar{\mathcal{E}}$ are positive.

Notice also that by definition, the word $\alpha_{\tau\pm 1}(a_i(z))$ is completely determined by its x-letter. Suppose that $|p| \leq 1$. This means that the two x-letters in the relations corresponding to the cells π_1, π_1' cancel. That implies, by Lemma 6.1, that the labels of the start edges of $\mathcal{B}_1, \mathcal{B}_2$ must be mutually inverse which contradicts the assumption that $\phi(\mathbf{top}(\mathcal{T}_1))$ is a reduced word. Hence $|p| = 2$, $\phi(p) = x_1 y_1^{-1}$, $x, y \in \mathcal{X}^{\pm 1}$, $x_1 \neq y_1^{\pm 1}$. Similarly if $\mathcal{T}_2$ corresponds to a rule from $\mathcal{S}$, $|q| = 2$, $\phi(q) = x_2 y_2^{-1}$, $x_2, y_2 \in \mathcal{X}^{\pm 1}$, $x_2 \neq y_2^{\pm 1}$.

Let $U_1 \equiv \phi(u_1)$, $U_2 \equiv \phi(u_2)$. Since Γ contains no cells (by Lemma 3.11), the equality $U_1 \phi(q) U_2 \phi(p)^{-1} = 1$ must be true in the free group.

Case 1. Suppose that $\mathcal{T}_2$ corresponds to a rule from $\bar{\mathcal{S}}$. Then

$$U_1 U_2 y_1 x_1^{-1} = 1$$

in the free group where U_1, U_2 are reduced products of fourth powers of letters from $\mathcal{X}^{\pm 1}$. Considering the homomorphism of the free group onto $\mathbb{Z}$ which kills all letters except x_1, we immediately get a contradiction.

Case 2. Suppose now that $\mathcal{T}_2$ corresponds to a rule from $\mathcal{S}$. Then the equality $U_1 x_2 y_2^{-1} U_2 y_1 x_1^{-1} = 1$ is true in the free group. Taking the projections on $\langle x_1 \rangle$, $\langle x_2 \rangle$ as in Case 1, we deduce that $x_1 \equiv x_2, y_1 \equiv y_2$.

Case 2.1. Suppose that U_1, U_2 are not empty. Then Lemma 6.10 (1) immediately gives a contradiction.

Case 2.2. Suppose that one of the words U_1 or U_2 is empty. Then the other word must be empty as well (these words are freely conjugate and reduced). Notice that the paths u_1, u_2 cannot contain consecutive edges with mutually inverse edges because otherwise the cells containing these edges cancel (they correspond to relations of the form (2.8) with the same a-letter). Hence the paths u_1 and u_2 are empty. Therefore the cells π_1 and π_2 have a common x-edge and a common a-edge. Since these two cells do not cancel (Δ is reduced), one of them corresponds to a relation of the form 2.6 and the other one corresponds to a relation of the form 2.7. (If these cells are both k-cells, then it is clear from the structure of the a-bands between π_1 and π_2, that $\mathcal{B}_1$ determines equal occurrences of a-letters in the boundary labels of π_1 and π_2.) Without loss of generality we can assume that π_1 corresponds to a relation of the form (2.6) and π_2 corresponds to a relation of the form (2.7). The Θ-edges on these cells have the same labels (up to the direction) because the label of the Θ-edge is encoded in the x-edges of a relation (2.6) or (2.7). Hence $\mathcal{T}_1$ and $\mathcal{T}_2$ correspond to mutually inverse rules of $\mathcal{S}$.

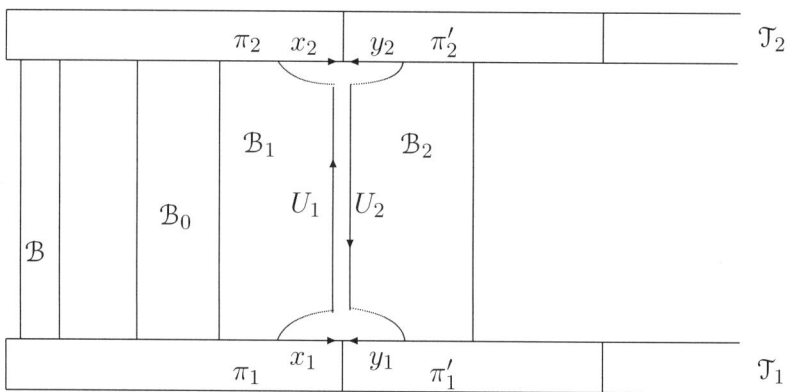

Figure 19.

Let $\mathcal{B}$ be the k-band starting on the k-edge of the cell π_1 (that belongs to $\mathbf{top}(\mathcal{T}_1)$) and ends on a k-edge of $\mathbf{bot}(\mathcal{T}_2)$. Assume that there is an a-band $\mathcal{B}_0$ connecting $\mathcal{T}_1$ and $\mathcal{T}_2$ between $\mathcal{B}$ and $\mathcal{B}_1$. Assume that it is the closest one to $\mathcal{B}_1$. Notice that it starts with an edge of $\mathbf{top}(\mathcal{T}_1)$ labelled by a negative a-letter, since the all a-labels of a-edges of π_1, situated between a k-edge and a θ-edge, are positive or negative simultaneously.

Consider the subdiagram Γ^0 between $\mathcal{B}_0$ and $\mathcal{B}_1$. Its contour has decomposition $u_1^0 q^0 u_2^0 (p^0)^{-1}$ similar to the decomposition $u_1 q u_2 p^{-1}$ of the contour of Γ. Now we obtain the equality $U_1^0 x U_2^0 x^{-1} = 1$ in the free group for the boundary label of Γ^0. Here U_2^0 must be empty since the word U_1 was empty. Hence U_1^0 is empty being the product of fourth powers of x-letters. There are fewer maximal a-bands between $\mathcal{B}$ and $\mathcal{B}_0$ than between $\mathcal{B}$ and $\mathcal{B}_1$. Therefore, arguing in this way, we finally conclude that $\phi(\mathbf{bot}(\mathcal{B}))$ is empty too. But then π_1 and π_2 must have a common k-edge, and they cancel because $\mathcal{T}_1$ and $\mathcal{T}_2$ correspond to mutual inverse rules of $\mathcal{S}$. This contradicts the fact that Δ is reducible. $\square$

If the base of a trapezium Δ is a 2-letter word, and its height is equal to 2, then Δ is called a *small* trapezium.

Lemma 6.12. *Let Δ be a small trapezium of the first or mixed type with reduced history. Then the projection of the words $V_1 = \phi(\mathbf{top}(\mathcal{T}_1))$, $V_2 = \phi(\mathbf{bot}(\mathcal{T}_2))$ onto $\mathcal{A} \cup \bar{\mathcal{A}} \cup \mathcal{K}$ are admissible words for $\mathcal{S}$.*

Proof. By Lemma 6.9, we need only check the positivity conditions of the definition of admissible words. Also it is clear that it is enough to prove the statement for V_1. Indeed, by Lemma 3.11 every a- and k-band starting on $\mathbf{top}(\mathcal{T}_1)$ ends on $\mathbf{bot}(\mathcal{T}_2)$ and vice versa; so the projections of V_1 and V_2 onto

$\mathcal{A} \cup \bar{\mathcal{A}} \cup \mathcal{K}$ are the same in the free group (in particular they do not contain $\bar{\mathcal{A}}$-letters).

Hence it is enough to consider the cases when $\mathfrak{T}_1$ corresponds to a rule from $\mathcal{S}$, V_1 is a zL_j-, zP_j- or $R_j z$-sector. All three cases are similar, so we consider only one of them when V_1 is a zL_j-sector. We need to show that the projection of V_1 onto $\mathcal{A}$ is positive.

Suppose that V_1 contains a negative letter a^{-1}, $a \in \mathcal{A}(\overleftarrow{L_j})$. First let a^{-1} be the last $\mathcal{A}$-letter of V_1. Since V_1 and V_2 are reduced, every a-band starting on $\mathbf{top}(\mathfrak{T}_1)$ ends on $\mathbf{bot}(\mathfrak{T}_2)$ and vice versa. In particular the a-band $\mathcal{C}$ starting on the last $\mathcal{A}$-edge of $\mathbf{top}(\mathfrak{T}_1)$ ends on the last $\mathcal{A}$-edge of $\mathbf{bot}(\mathfrak{T}_2)$, hence the last $\mathcal{A}$-letter of V_2 is also a^{-1}.

Relations (2.6), (2.7) show that then V_1 ends with $a^{-1}x$ where $x = x(b, \tau)^{\pm 1}$, and V_2 ends with $a^{-1}x'$ where $x' = x(c, \tau')^{\pm 1}$, for some $b, c \in \mathcal{A}$. Let Γ be the diagram bounded by $\mathbf{top}(\mathcal{C})$, $\mathbf{bot}(\mathcal{B}')$, a portion of $\mathbf{top}(\mathfrak{T}_1)$ and a portion of $\mathbf{bot}(\mathfrak{T}_2)$. By Lemma 3.11 and 6.5, Γ contains no cells. Both words $\phi(\mathbf{top}(\mathcal{C}))$ and $\phi(\mathbf{bot}(\mathcal{B}'))$ are products of fourth powers of letters from $\mathfrak{X}$, and x and x' are not mutually inverse because $\tau' \neq \tau^{-1}$ by an assumption of the lemma. This immediately leads to a contradiction by Lemma 6.10 (2).

Thus the last a-letter of V_1 (resp. V_2) is positive. Since V_1 contains a negative a-letter a^{-1}, $(V_1)_a$ must contain a subword of the form $a^{-1}b$ for some $b \in \mathcal{A}$. But this contradicts Lemma 6.11. □

The next lemma shows how to get rid of $\mathcal{G}$-cells in some trapezia.

Lemma 6.13. *Let Δ be a quasitrapezium of the first type. Suppose that the history of Δ does not contain $\tau \in \mathcal{S}(34) \cup \mathcal{S}(4) \cup \mathcal{S}(45)$ or the base of Δ is empty. Then there exists a quasitrapezium Δ' of the first type such that*

- *Δ' has the same labels for its bottom, left and right sides as Δ,*

- *$\phi(\mathbf{top}(\Delta)) = \phi(\mathbf{top}(\Delta'))$ modulo $\mathcal{G}$-relations*

- *Δ' has no $\mathcal{G}$-cells and the type of Δ' is not higher than that of Δ.*

Proof. We only consider the case when Δ is a trapezium with non-empty base. The other cases are similar.

If Δ does not contain $\mathcal{G}$-cells then we can take $\Delta' = \Delta$, so suppose that Δ contain $\mathcal{G}$-cells.

Suppose there is a small subtrapezium Γ in Δ with base yz where yz is one of the words $(P_1 R_1)^{\pm 1}$, $P_1 P_1^{-1}$, $R_1^{-1} R_1$. Assume that Γ is crossed by maximal θ-bands $\mathfrak{T}_t$ and $\mathfrak{T}_{t+1}$. Denote $q_1 = \mathbf{bot}(\mathfrak{T}_t),...,q_4 = \mathbf{top}(\mathfrak{T}_{t+1})$. Let $V_i = \phi(q_i)$, $i = 1, ..., 4$. Notice that V_2, V_3, V_4 are words over $\mathcal{A}(P_1)$ by Lemmas 6.1, 6.3 and $V_2 = V_3$ modulo $\mathcal{G}$-relations by Lemma 6.6.

Denote by Γ_0 the subdiagram of Γ with the boundary label $V \equiv V_2 V_3^{-1}$. Assume Γ_0 contains at least one $\mathcal{G}$-cell. Then by Lemma 3.9 it has exactly one $\mathcal{G}$-cell and no other cells.

Let τ be the history of $\mathcal{T}_{t+1}$. We cut $\mathcal{T}_{t+1}$ along a $\theta(\tau, P_1)$-edge and the boundary of Γ_0 to obtain a closed subpath p labelled by the word $\theta(\tau, P_1)^{-1} V \theta(\tau, P_1)$. It is freely equal to $\theta(\tau, P_1)^{-1} V \theta(\tau, P_1) V^{-1} V$. Removing the subdiagram bounded by this loop, we get a hole in the diagram Δ.

Recall that $\theta(\tau, P_1)$ commutes with $\mathcal{A}(P_1)$-letters by relations (2.6), since τ does not belong to $\mathcal{S}(34) \cup \mathcal{S}(4) \cup \mathcal{S}(45)$ by the assumptions of the lemma.

Hence there exists a θ-band with boundary label $\theta(\tau, P_1)^{-1} V \theta(\tau, P_1) V^{-1}$. If we connect a $\mathcal{G}$-cell by a vertex to the end of the first θ-cell of this band, we get a diagram Γ_1 with boundary label freely equal to $\phi(p)$. Now we can fill the hole bounded by p by the diagram Γ_1. As a result the θ-bands of Δ, except $\mathcal{T}_{t+1}$ do not change, the band $\mathcal{T}_{t+1}$ gets longer, and the $\mathcal{G}$-cell moves outside Γ toward the top of Δ. Let $\tilde{\Delta}$ be the resulting diagram, and Δ_1 be the diagram obtained by reducing $\tilde{\Delta}$.

Two k-cells of the same k- or θ-band in $\tilde{\Delta}$ cannot form a reducible pair of cells (see (ii) in the definition of non-reduced diagram) because otherwise the diagram Δ would not be reduced. Hence k-cells are not removed in the process of reducing $\tilde{\Delta}$.

This implies that Δ_1 satisfies the definition of a trapezium. The height of Δ_1 is the same as the height of Δ, the diagram between $\mathcal{T}_1$ and $\mathcal{T}_{t+1}$ in Δ_1 contains fewer $\mathcal{G}$-cells than the corresponding subdiagram of Δ. Hence after a number of such transformations, we get a trapezium Δ_s with the same boundary label as Δ, and all $\mathcal{G}$-cells in Δ_s are between $\mathbf{top}(\mathcal{T}_d)$ and the top path p_2. Notice that Δ_s has the same number of k- and $\mathcal{G}$-cells as Δ but possibly a bigger number of auxiliary (θ, a)-cells.

For every $\mathcal{G}$-cell π in Δ_s, there exists a path in Δ_s which does not cross any θ-bands and connects π with p_2. Hence all $\mathcal{G}$-cells in Δ_s can be cut off from Δ_s without changing the bottom, left and right sides of Δ_s. The label of the top side does not change modulo $\mathcal{G}$-relations. After we remove all $\mathcal{G}$-cells from Δ_s, we get a trapezium Δ' satisfying the first two of the required conditions. Since Δ contains $\mathcal{G}$-cells, the type of Δ' is smaller than the type of Δ: to obtain Δ' from Δ, we add auxiliary θ-cells and remove $\mathcal{G}$-cells. □

Lemma 6.14. *For every quasiring Δ with empty base and contours p and q there exists a quasiring with the same boundary labels and a recursively bounded number of cells.*

Proof. By Lemma 5.6 we can assume that Δ satisfies condition (R1).

By the definition, there is an $a(z)$-band, connecting p and q, in Δ. Since the base of Δ is empty, every θ-cell of Δ must be an $\mathcal{A}(z)$-cell, and $z \neq P_1$ by the definition of a quasiring.

Denote by Δ' the minimal annular subdiagram of Δ which contains all the θ-annuli of Δ. It contains no $\mathcal{G}$-cells by Lemma 6.5(2). Hence every a-band starting on p ends on q and vice versa. Therefore the length of an arbitrary θ-annulus is linearly bounded in terms of $|p|, |q|$, and we have a bounded number of possibilities for their labels. Since Δ is reduced, we conclude that the number of its θ-annuli is also recursively bounded. The statement is obtained after the application of Lemma 4.4 to the remaining annular subdiagrams of Δ which are diagrams over $\mathcal{H}_1$.

$\square$

The following lemma shows that trapezia of the first type simulate the work of $\mathcal{S}$. Notice that this lemma does not follow from Lemma 6.9 because admissible words for $\mathcal{S}$ differ from admissible words for $\mathcal{S} \cup \bar{\mathcal{S}}$.

Lemma 6.15. *Let Δ be a trapezium of the first type with base B and the reduced history $h = \tau_1 \cdot \ldots \cdot \tau_d$, $\tau_i \in \mathcal{S}$, $d \geq 2$, with θ-bands $\mathcal{T}_1, \ldots, \mathcal{T}_d$. Let W (resp. W') be the projection of the label of the bottom (resp. top) path of Δ onto $\mathcal{A} \cup \bar{\mathcal{A}} \cup \mathcal{K} \cup \bar{\mathcal{K}}$, U_i (resp. V_i), $i = 1, \ldots, d$, be the projection of $\phi(\mathbf{bot}(\mathcal{T}_i))$ (resp. $\phi(\mathbf{top}(\mathcal{T}_i))$) on the same set. Then $U_2, \ldots, U_d, V_1, \ldots, V_{d-1}$ are admissible words for $\mathcal{S}$. In addition*

$$W = U_1 (\mathrm{mod}\ \mathcal{G}), ||V_1| - |U_1|| \leq c|B|, U_2 = V_1 (\mathrm{mod}\ \mathcal{G}), V_2 = U_2 \circ \tau_2,$$
$$||V_2| - |U_2|| \leq c|B|, \ldots, V_{d-1} = U_{d-1} \circ \tau_{d-1}, ||V_{d-1}| - |U_{d-1}|| \leq c|B|,$$
$$U_d = V_{d-1} (\mathrm{mod}\ \mathcal{G}), ||V_d| - |U_d|| \leq c|B|, W' = V_d (\mathrm{mod}\ \mathcal{G})$$

where c is the maximum of lengths of words in $\bar{\mathcal{E}}$. If Δ does not contain $\mathcal{G}$-cells then one can remove "(mod $\mathcal{G}$)" from the previous statement (i.e. all equalities will be true in the free group).

Proof. The proof is similar to the proof of Lemma 6.9. The fact that $U_2, \ldots, U_d, V_1, \ldots, V_{d-1}$ are admissible words for $\mathcal{S}$ follows from Lemma 6.12.

$\square$

The definition of admissible words for $\bar{\mathcal{S}}$ does not have the positivity conditions. This makes analysis of computations of $\bar{\mathcal{S}}$ different from the analysis of computations of $\mathcal{S}$. On the one hand the analysis of computations of $\bar{\mathcal{S}}$ is more difficult because there are more possible bases of admissible words for $\bar{\mathcal{S}}$. On the other hand the following lemma which is an immediate corollary of Lemma 3.3, part 2, allows us to replace any computation of the form $U \bullet h = V$ of $\bar{\mathcal{S}}$ by any other computation with the same initial and terminal words. In the case of computations over $\mathcal{S}$ the possibilities of modifying the computation were very limited: we could only replace h by a reduced form of h or we could remove certain subwords of h (which fix certain admissible words).

We call a computation $U \bullet h = V$ *reduced* if the word h is freely reduced.

Lemma 6.16. *For every reduced computation $U \bullet h = V$ of $\bar{\mathcal{S}}$ there exists a trapezium with top label V, bottom label U, and history h. If $U = U'z$, $V = V'z_1$ where z (resp. z_1) coincides with the first letter of U' (resp. V') then for every computation $U \bullet h = V$ of $\bar{\mathcal{S}}$ there exists a ring of type 2 with boundary labels U' and V'.*

Proof. Assume that $|h| = 1$, and $h = \tau \in \mathcal{S}^+$. We consider a path p on the plane labelled by U. To every a-edge of p we attach a cell corresponding to an appropriate $\bar{\Theta}$-analog of relation (2.7) which in turn corresponds to the rule τ. Similarly, to every k-edge of p we attach a cell corresponding to (2.6). Then the $\bar{\Theta}$-edges of the neighbor cells should be identified. We get a $\bar{\Theta}$-band $\mathcal{T}$. (Of course we identify some other labels, if necessary, to obtain a reduced label for $\mathbf{top}(\mathcal{T})$.) The form of these relations shows that we obtain a band $\mathcal{T}$ with top label graphically equal to V.

Two k-cells of $\mathcal{T}$ cannot form reducible pair of k-cells since the admissible words U and V have no subwords $z^{\pm 1}z^{\mp 1}$ for a k-letter z. Hence $\mathcal{T}$ is a reduced diagram.

Similarly, one can construct $\mathcal{T}$ if $\tau \in \mathcal{S}^-$.

If $|h| = d > 1$, then we construct the diagram Δ by gluing together d bands $\mathcal{T}_1, \ldots, \mathcal{T}_n$, corresponding to the letters of the history h. Two cells from distinct bands cannot form a reducible pair since the history h is reduced. Hence Δ is a reduced diagram, and it is the desired trapezium.

To prove the second statement, we notice that in this case the resulting trapezium is bounded by two k-bands which are copies of each other. Hence we can identify them to get the desired ring (or if the resulting annular diagram is compressible, then there is a ring of smaller type) with the same boundary labels. However we should verify that the words U', V' are not conjugates in $\mathcal{H}_1$ of words having fewer occurrences of k-letters. Assume there is a reduced annular diagram for such a conjugation with contours p and q where $\phi(p) \equiv U$. Then there must be a k-band $\mathcal{B}$ which starts and ends on p such that there are no k-bands in the simply connected subdiagram Γ bounded by $\mathcal{B}$ and p. The subdiagram Γ contains no θ-edges by Lemma 3.11 since $\phi(p)$ has no θ-letters. Therefore the word u written between corresponding mutual inverse k-letters of U is equal to a word having no a-letters. Hence if $|u| > 0$, then there must be an a-band in Γ, which starts and ends on p, and gives a cancellation in u. If $|u| = 0$ then the k-letters cancel. In both cases we get a contradiction to the fact that the admissible word U must be reduced. □

7 Computations of $\mathcal{S} \cup \bar{\mathcal{S}}$

Notice that if we cut a ring along the top side of a k-band $\mathcal{B}$, we get a rectangular van Kampen diagram. If we then attach a copy of $\mathcal{B}$ along its top side to one of the sides of the rectangle, we get a trapezium Δ with boundary $p_1p_2p_3p_4$ which satisfies the additional property that p_1, p_3^{-1} are sides of two copies of the same k-band (the band $\mathcal{B}$). The top and bottom sides of Δ are the outer and inner boundary components of the initial ring.

As we know (see Lemma 6.9), with every trapezium Δ with bottom label U and top label V, we can associate a computation $U_{\setminus x} = W_0, W_1 = W_0 \circ \sigma_1, ..., W_n = W_{n-1} \circ \sigma_n = V_{\setminus x}$ of the machine $\mathcal{S} \cup \bar{\mathcal{S}}$ (modulo $\mathcal{G}$ if the trapezium contains $\mathcal{G}$-cells). Here $U_{\setminus x}$ and $V_{\setminus x}$ are obtained from U and V by deleting their x-letters. The words W_i are the $\mathcal{A} \cup \bar{\mathcal{A}} \cup \mathcal{K} \cup \bar{\mathcal{K}}$-projections of the labels of sides of θ-bands in Δ.

Computations modulo $\mathcal{G}$ will be called $\mathcal{G}$-*computations*. Previously considered computations will be called *free*. The concept of admissible word W is defined naturally for $\mathcal{G}$-computations: we add to the reducibility of such a word that W has subwords neither of the form $P_1 u P_1^{-1}$ nor of the form $R_1^{-1} u R_1$ for a word u in $\mathcal{A}(P_1)$-letters which is equal to 1 modulo $\mathcal{G}$-relations. Locked sectors are defined naturally for $\mathcal{G}$-computations as well (if a rule locks a zz_+-sector then all admissible words in the domain of the rule must have zz_+-sectors with inner parts empty modulo $\mathcal{G}$-relations). Notice that $\mathcal{G}$-computations can be viewed as computations of S-machines obtained from $\mathcal{S} \cup \bar{\mathcal{S}}$ by replacing the free product of $\mathcal{G}$ and the free group over $\{a_{m+1}(P_1), \ldots, a_{\bar{m}}(P_1)\}$ for the free subgroup generated by $\mathcal{A}(P_1)$ in the hardware.

Notice that if $U^{\pm 1}$ does not contain $P_1 z$- or zR_1-sectors then free computations and computations modulo $\mathcal{G}$ are the same. Also any computation of $\bar{\mathcal{S}}$ is free.

Given a ring Δ, by Lemma 6.15, we get a $\mathcal{G}$-computation $U_1 \bullet \sigma_1 ... \sigma_d = V_d$. Suppose that we have recursively bounded the number d in terms of the lengths of the contours of the diagram. If the base of the ring contains P_j or R_j for some j then we can cut the ring along a side of the corresponding k-band which does not contain x-edges. This side is of bounded length, and the resulting diagram will be simply connected. Now using Lemma 5.1, we can recursively bound the number of cells in a ring.

If the base does not contain P_j and R_j for any j then, to bound the number of cells in Δ from above, we can assume by lemmas 6.4 and 6.5 that the ring does not contain $\mathcal{G}$-cells. Therefore the words U_i, V_i are of recursively bounded lengths. By Lemma 4.4, the number of cells between any two consecutive θ-annuli in Δ is bounded.

In both cases the history of the the resulting ring coincides with the history of Δ because it cannot be shortened, by condition (iv) in the definition of a reduced diagram Δ. Thus we obtain

Lemma 7.1. *If the length of the computation corresponding to a ring Δ is recursively bounded, then there exists a ring with the same boundary labels and the same history as Δ whose number of cells is recursively bounded in terms of lengths of boundaries of Δ.*

If a computation (a $\mathcal{G}$-computation) corresponds to a ring we call it a *free ring computation* (a *ring computation*). A ring computation is a reduced computation by Lemma 6.8.

For every word h and every natural number $t \le |h|$ we denote the prefix of h of length t by $h[t]$.

Lemma 7.2. *If $U \bullet h = V$ is a ring computation then $U \bullet h[t]$ is also a ring computation for every $t = 1, ..., |h|$.*

Proof. The computation $U \bullet h[t]$ corresponds to a subring of the ring corresponding to $U \bullet h = V$ containing the first t θ-bands of that ring. $\square$

By definition the *base of a computation* is the base of the starting word U (which is the same as the base of V), and the *history of computation* $U \bullet h$ is h. This agrees with the definitions of the base and the history of a trapezium.

Remark 7.1. When we cut a ring to make a trapezium, we can choose any of the k-bands of the ring. Thus any letter of the base of the ring can be chosen to be the first letter (and the last letter) of the base of the computation. Also by taking a mirror image of a trapezium, we get a trapezium with inverse base. For simplicity let us always assume that if the base of a ring computation contains a letter $K_1^{\pm 1}$, then the base starts and ends with K_1.

Here is a translation of Lemma 6.13 into the language of computations. We leave it as an exercise to the reader to translate the proof of Lemma 6.13 as well.

Lemma 7.3. *Let $U \bullet h = V$ be a ring computation of $\mathcal{S}$. Suppose that h does not contain letters from $\mathcal{S}(34) \cup \mathcal{S}(4) \cup \mathcal{S}(45)$. Then there exists a free ring computation $U \bullet h = V'$ where $V = V'$ modulo $\mathcal{G}$.*

7.1 Brief history

Lemma 7.4. *Let $U \bullet h = V$ be a reduced free computation of $\mathcal{S}(i)$ or of $\bar{\mathcal{S}}(i)$, $i = 1, ..., 5$, with $|h| = 2$, $h = \sigma_1 \sigma_2$. Suppose that U is a zz_+-sector. Then inequalities $|U| < |U \circ \sigma_1| > |V|$ cannot happen simultaneously.*

Proof. Indeed, the definition of $\mathcal{S} \cup \bar{\mathcal{S}}$ shows that if $|U| < |U \circ \sigma_1|$, $\sigma_1 \in \mathcal{S} \cup \bar{\mathcal{S}}$, then, depending on g, $(U \circ \sigma_1)_a = U_a b$, or $(U \circ \sigma)_a = bU_a$, $b \in \mathcal{A} \cup \bar{\mathcal{A}}$, where $U_a b$ (resp. bU_a) is a freely reduced word, and σ_1 is completely determined (by Lemma 2.6) among the rules from $\mathcal{S}(i)$ by the letter b. If in addition $|V| < |U \circ \sigma_1|$, then $V_a = U_a b b^{-1}$ (resp. $V_a = b^{-1} b U_a$) since $V = (U \circ \sigma_1) \circ \sigma_2$. Therefore $\sigma_2 = \sigma_1^{-1}$, a contradiction with the assumption that the computation $U \bullet h = V$ is reduced. $\square$

Lemma 7.4 immediately implies the following

Lemma 7.5. *Let $U \bullet h = V$ be a reduced free computation of $\mathcal{S}(i)$ or of $\bar{\mathcal{S}}(i)$, $i = 1, \ldots, 5$, and U is a zz_+-sector. Suppose that rules of $\mathcal{S}(i)$ are active with respect to zz_+-sectors. Then the length $|h|$ is at most $|U| + |V|$.*

Let $U \bullet h = V$ be a computation. A maximal nonempty subword of h consisting of rules from $\mathcal{S}(\omega)$ for a given $\omega \in \{1, 2, 3, 4, 5\}$ will be called an *age* of ω in the computation. The word h has a unique decomposition $g_1 h_1 g_2 h_2 \ldots$ where g_i are transition rules (g_1 may be missing) and h_i are ages ($i = 1, 2, \ldots$). If $g_i \in \mathcal{S}(\omega'_i)$, h_i is an age of ω_i, $i = 1, 2, \ldots$, then the word $(\omega'_1)(\omega_1)(\omega'_2)(\omega_2)\ldots$ will be called the *brief history* of the computation $U \bullet h = V$ and is denoted by $br(h)$.

Lemma 7.6. *Let $U \bullet h = V$ be a free reduced computation of $\mathcal{S}$ or $\bar{\mathcal{S}}$. Suppose that U is a zz_+-sector, and for some $\omega, \omega' \in \{1, 12, \ldots, (5), (51)\}$, rules from $\mathcal{S}(\omega)$ lock zz_+-sectors and rules from $\mathcal{S}(\omega')$ are active with respect to zz_+-sectors. Then the brief history $br(h)$ does not contain subwords $(\omega)(\omega')(\omega)$.*

Proof. Suppose that $br(h) = \beta_1(\omega)(\omega')(\omega)\beta_2$ for some prefix β_1 and some suffix β_2. Let $h = h_1 h_2 h_3 h_4 h_5$ be the corresponding decomposition of h. Since rules from $\mathcal{S}(\omega)$ lock zz_+-sectors, we have that the words $(U \circ h_1 h_2)_a$ and $(U \circ h_1 h_2 h_3 h_4)_a$ are empty. Since rules from h_3 are active with respect to zz_+-sectors, and are from the same $\mathcal{S}(\omega')$, there exists a word u in $\mathcal{A} \cup \bar{\mathcal{A}}$ such that $(U \circ h_1 h_2 \sigma_3)_a$ is freely equal to either $(U \circ h_1 h_2)_a u$ or $u(U \circ h_1 h_2)_a$. Hence u is freely equal to 1. Since by Lemma 2.6 each letter in u and ω' uniquely determines a rule in $\mathcal{S}$ (or $\bar{\mathcal{S}}$), we conclude that h_3 is not freely reduced or empty, a contradiction. $\square$

Consider a brief history of the form $h \equiv (12)(2)(23)(3)(34)(4)(45)(5)(51)$. Any word of the form $h^{\pm 1}$ will be called a *historical period*. (We set here $(1)^{-1} = (1), (12)^{-1} = (12), \ldots$, though one can prefer the formula $(12)^{-1} = (21)$.)

Lemma 7.7. *Suppose that the base of a free reduced computation $U \bullet h = V$ of $\mathcal{S}$ or $\bar{\mathcal{S}}$ contains a subword $\overleftarrow{L_j} L_j P_j R_j \overrightarrow{R_j}$ for $j \neq 1$. Then $br(h)$ is a subword of a word of the form*

$$f_0 h_1 f_1 \ldots h_s f_s$$

where $h_1, \ldots, h_s$ are historical periods and each of (possibly empty) words $f_0, \ldots, f_s$ are ages of (1).

Proof. Indeed, notice that for every $i = 2, 3, 4, 5$ rules in $\mathcal{S}(i)$ are active with respect to zz_+-sectors for at least two letters z occurring in $\overleftarrow{L_j} L_j P_j R_j \overrightarrow{R_j}$, and each rule from $\mathcal{S}(ii') \cup \bar{\mathcal{S}}(ii') \cup \mathcal{S}(i'i) \cup \bar{\mathcal{S}}(i'i)$ locks one of these two sectors. Now using Lemma 7.6, we can conclude that $br(h)$ does not contain subwords of the form $(i'i)(i)(i'i)$ for any i. This immediately implies that $\mathrm{br}(h)$ has the desired form. □

Lemma 7.8. *Let $U \bullet h = V$ be a ring computation of $\mathcal{S}$ (and the first and the last letters of the base B of U the same). Then $\mathrm{br}(h)$ does not contain subwords $(12)(2)(12)$, $(23)(2)(23)$, $(23)(3)(23)$, $(34)(4)(34)$, $(45)(4)(45)$, and $(51)(5)(51)$.*

Proof. Notice that since the first and the last letters in the base B are the same, for every letter z occurring in B, U contains a zz'-sector (for some z').

Let $\mathrm{br}(h)$ contain a subword $(12)(2)(12)$. First assume that the base B contains L_j or $\overleftarrow{L}_j$ for some j. Since $\mathrm{br}(h)$ contains (12) and rules from $\mathcal{S}(12)$ lock zL_j-sectors, we can conclude that B contains a subword $\overleftarrow{L}_j L_j$ (here we use the assumption that the first and the last letters in B are the same, otherwise we could have L_j as the first letter in B, and B could contain no copies of $\overleftarrow{L}_j L_j$). Since rules of $\mathcal{S}(2)$ are active with respect to $\overleftarrow{L}_j L_j$-sectors, we get a contradiction with Lemma 7.6. Therefore we can assume that B contains neither L_j nor $\overleftarrow{L}_j$. By passing to a subcomputation, we can assume that $\mathrm{br}(h) = (12)(2)(12)$.

Notice that the rules from $\mathcal{S}(12) \cup \mathcal{S}(2)$ do not change modulo $\mathcal{G}$ the a-projections of admissible words without zL_j- and $L_j z$-sectors. Hence the a-projections of corresponding sectors in U and V are the same modulo $\mathcal{G}$. The $\bar{\mathcal{E}}$ and Ω-coordinates of U and V are $(1, \emptyset)$. Hence U and V are equal modulo $\mathcal{G}$, and the labels of the corresponding θ-annuli of the ring are equal modulo $\mathcal{G}$-relations too by Lemma 6.1. Hence the ring is not reduced (see condition (iv) in the definition of reduced diagram), a contradiction.

The other five statements can be proved by contradiction quite analogously. □

7.2 Standard computations

A computation $U \bullet h = V$ of $\mathcal{S}$ (of $\bar{\mathcal{S}}$) is said to be *standard* or *j-standard* computation if its base has the form $(\overleftarrow{L}_j L_j P_j R_j \overrightarrow{R}_j)^{\pm 1}$ and the computation is reduced. If U is an admissible word with this base then for each $z \in \{\overleftarrow{L}_j, L_j, P_j, R_j\}$ let $z(U)$ be the a-projection of the zz_+-sector in U.

Lemma 7.9. *Let $U \bullet h = V$ be a reduced free computation of $\mathcal{S}(i)$, $i \in \{2, 3, 4, 5\}$. Suppose that the base of the computation is $\overleftarrow{L}_j L_j P_j$,*
(1) The equality

$$|\overleftarrow{L}_j(U)| + |L_j(U)| = |\overleftarrow{L}_j(U \circ h[t])| + |L_j(U \circ h[t])|$$

holds for every $t = 1, ..., |h|$.
(2) Let $z \in \{\overleftarrow{L}_j, L_j\}$. Then the sequence of lengths $|z(U \circ h[t])|$, $t = 0, 1, ...$ is monotone.

THE CONJUGACY PROBLEM AND HIGMAN EMBEDDINGS 65

Proof. Indeed, since all words $U \circ h[t]$ are admissible for $\mathcal{S}$, the words $\overleftarrow{L}(U \circ h[t])$ and $L_j(U \circ h[t])$ are positive for $t = 1, ..., |h|$. If $i \neq 2, 4$ then the rules $\tau \in \mathcal{S}(i)$ are not active with respect to $\overleftarrow{L}_j L_j$- and $L_j P_j$-sectors. Hence in that case the words $U \circ h[t]$ are all the same ($t = 1, ..., |h|$) which proves (1) and (2).

If $i \in \{2, 4\}$ then by Lemma 2.6 for every rule $\tau \in \mathcal{S}(i)$ there exist two a-letters $b_1 = b_1(\tau), b_2 = b_2(\tau)$ such that for every admissible word W, with base $\overleftarrow{L}_j L_j P_j$, $\overleftarrow{L}_j(W \circ \tau) = \overleftarrow{L}_j(W) b_1$, $L_j(W \circ \tau) = b_2 L_j(W)$ and the rule τ is completely determined by each of the letters b_1, b_2. Moreover b_1 is a positive letter if and only if b_2 is a negative letter. Since the words $\overleftarrow{L}_j(U \circ h[t])$, $L_j(U \circ h[t])$ are always positive and h is reduced, we see that either for all τ in h the letter $b_1(\tau)$ is positive and the letter $b_2(\tau)$ is negative, or for all letters τ of h, $b_1(\tau)$ is negative and $b_2(\tau)$ is positive. Therefore either for every $t = 1, ..., |h|-1$, $|\overleftarrow{L}_j(U \circ h[t+1])| = |\overleftarrow{L}(U \circ h[t])|+1$ and $|L_j(U \circ h[t+1])| = |L_j(U \circ h[t])|-1$ or for every $t = 1, ..., |h|-1$, $|\overleftarrow{L}_j(U \circ h[t+1])| = |\overleftarrow{L}(U \circ h[t])|-1$ and $|L_j(U \circ h[t+1])| = |L_j(U \circ h[t])|+1$. This implies parts (1) and (2) again. $\square$

For every word W, we denote by $\operatorname{diff}(W)$ the difference between the number of positive, and the number of negative occurrences of a-letters in a word W. Also let, as before, c be the maximal length of relations in $\overline{\mathcal{E}}$.

Let h be a word over $\mathcal{S}$. For every $t \leq |h|$ let $s(t)$ be the number of occurrences of rules from $\mathcal{S}(34)$ in $h[t]$.

Lemma 7.10. *Let $U \bullet h = V$ be a j-standard computation of $\mathcal{S}$, $j \neq 1$. Suppose that for some $z \in \{\overleftarrow{L}_j, L_j, R_j\}$, and some $t \leq |h|$*

$$|z(U \circ h[t])| > s(t)c + |U|. \tag{7.1}$$

Then, for every $d \geq t$,

(a) the d-th rule of h does not belong to $\mathcal{S}(12) \cup \mathcal{S}(1) \cup \mathcal{S}(34) \cup \mathcal{S}(51)$ if $z \in \{\overleftarrow{L}_j, R_j\}$ and the d-th rule of h does not belong to $\mathcal{S}(34)$ if $z = L_j$,

(b) if $z \in \{\overleftarrow{L}_j, R_j\}$ then the word $z(U \circ h[d])$ is not empty; if $z = L_j$ then at least one of the words $\overleftarrow{L}_j(U \circ h[d])$ or $L_j(U \circ h[d])$ is not empty,

(c) $|U \circ h[d]| \geq d - t$.

Proof We may assume that the base of U has the form $\overleftarrow{L}_j L_j P_j R_j \overrightarrow{R}_j$ because otherwise we can replace U by U^{-1}.

Case 1. Suppose that $z = \overleftarrow{L}_j$. Without loss of generality we can assume that t is the minimal number such that (7.1) holds. From (7.1), we have that $t > 0$.

Then the t-th rule of h belongs to $\mathcal{S}(2) \cup \mathcal{S}(34) \cup \mathcal{S}(4)$ because otherwise $\overleftarrow{L}_j(U \circ h[t-1]) = \overleftarrow{L}_j(U \circ h[t])$ which contradicts the minimality of t. Similarly this rule cannot belong to $\mathcal{S}(34)$ because in that case $s(t-1) = s(t) - 1$ and

$$|\overleftarrow{L}_j(U \circ h[t-1])| \geq |\overleftarrow{L}_j(U \circ h[t])| - c > s(t)c - c + |U| \geq s(t-1)c + |U|$$

contrary to the minimality of t.

Furthermore, the minimality of t and the form of the rules from $\mathcal{S}(2)\cup\mathcal{S}(4)$ imply that $|\overleftarrow{L}_j(U\circ h[t])|=|\overleftarrow{L}_j(U\circ h[t-1])|+1$.

Case 1.1. Suppose first that the t-th rule in h belongs to $\mathcal{S}(2)$. If for every $t'\geq t$ the t'-rule in h is from $\mathcal{S}(2)$ then

$$|\overleftarrow{L}_j(U\circ h[t])|<|\overleftarrow{L}_j(U\circ h[t+1])|<...<|\overleftarrow{L}_j(U\circ h)|$$

by Lemma 7.4. (The series increases because of the minimality of t.) This implies parts (a), (b), (c) of the conclusion of the lemma since $\mathcal{S}(12)$ locks $\overleftarrow{L}_j$-sector.

Suppose that for some $t'>t$ the t'-th rule in h is not from $\mathcal{S}(2)$. Let t' be the first such number. Then by Lemma 7.7 the t'-th rule in h belongs to $\mathcal{S}(12)\cup\mathcal{S}(23)$. Since rules from $\mathcal{S}(12)$ lock $\overleftarrow{L}_jL_j$-sectors, the t'-th rule in h cannot belong to $\mathcal{S}(12)$ by Lemma 7.4. So it belongs to $\mathcal{S}(23)$.

Clearly the claims of the lemma are true if $t'=|h|$.

Suppose that $t'<|h|$. Then by Lemma 7.7 $h\equiv h[t']h'h''$ where h' is a maximal subword of h consisting of rules from $\mathcal{S}(3)$. We will show that h'' is empty.

Indeed, suppose that h'' is not empty. Then by Lemma 7.7, h'' starts with a rule τ from $\mathcal{S}(34)$. Let $U'=U\circ h[t']h'$. Since τ locks P_jR_j-sectors, and since all words $U\circ h[u]$ are admissible for $\mathcal{S}$ ($u=0,1,...,|h|$), we have that $P_j(U')=\emptyset$ and the words $\overleftarrow{L}_j(U')$, $L_j(U')$ and $R_j(U')$ are positive.

Therefore, by Lemma 7.9,

$$\begin{aligned}\operatorname{diff}(U')=|U_a'|&\geq |\overleftarrow{L}_j(U')|=|\overleftarrow{L}_j(U\circ h[t'])|\\&\geq |\overleftarrow{L}_j(U\circ h[t])|+(t'-t)>s(t)c+|U_a|+(t'-t).\end{aligned} \quad (7.2)$$

But it is easy to see that for every rule τ in $\mathcal{S}\setminus\mathcal{S}(34)$ and for every admissible word W in the domain of τ, $\operatorname{diff}(W)=\operatorname{diff}(W\circ\tau)$, and for every $\tau\in\mathcal{S}(34)$, $\operatorname{diff}(W\circ\tau)\leq\operatorname{diff}(W)+c$. Since there are at most $s(t)$ rules from $\mathcal{S}(34)$ in $h[t']h'$, we can conclude that

$$\operatorname{diff}(U')\leq\operatorname{diff}(U)+s(t)c\leq|U_a|+s(t)c, \qquad (7.3)$$

a contradiction with (7.2).

Thus h'' is empty, and part (a) of the lemma is established. Since rules from $\mathcal{S}(3)$ are not active with respect to $\overleftarrow{L}_jL_j$-sectors, and $\overleftarrow{L}_j(U\circ h[t])$ is not empty, part (b) follows as well.

Recall that h' is not empty because $t'\neq|h|$. Notice that the $R_j\overrightarrow{R}_j$-sector W of $U\circ h[t']$ is in the domain of h'. By Lemma 7.5 applied to the computation $W\bullet h'$, we have

$$|h'|=|h|-t'\leq |R_j(W)|+|R_j(W\circ h')|=|R_j(U\circ h[t'])|+|R_j(U\circ h)|. \quad (7.4)$$

Notice that rules in $\mathcal{S}(23)$ lock $R_j\overrightarrow{R}_j$-sectors, so $|R_j(U\circ h[t'])|=0$ (since the last rule in $h[t]$ is from $\mathcal{S}(23)$), and therefore $R_j(U\circ h)|\geq d-t'$ by (7.4).

Notice also that $|\overleftarrow{L}_j(U \circ h[d])| \geq |\overleftarrow{L}_j(U \circ h[t])| + t' - t$ by (7.2) and the equality $\overleftarrow{L}_j(U \circ h[d]) = \overleftarrow{L}_j(U \circ h[t'])$. Hence for every $d \geq t'$

$$|U \circ h[d]| > |\overleftarrow{L}_j(U \circ h[d])| + |R_j(U \circ h[d])|$$
$$\geq (|\overleftarrow{L}_j(U \circ h[t])| + t' - t) + (d - t')$$
$$\geq s(t)c + d - t + |U| \geq d - t,$$

which proves part (c).

Case 1.2. The case when the t-th rule in h belongs to $\mathcal{S}(4)$ is similar. Let $h = h[t]h'$. Then the brief history of h' cannot start with $(4)(45)$ by Lemma 7.4 since rules from $\mathcal{S}(45)$ lock $\overleftarrow{L}_j L_j$-sectors.

The brief history also cannot start with $(4)(34)$. Indeed, suppose that $h' = h''\tau h'''$ where h'' is a word over $\mathcal{S}(4)$, $\tau \in \mathcal{S}(34)$. Then as in Case 1.1, we denote $U' = U \circ h[t]h''$. Then we can use the fact that τ locks $P_j R_j$-sectors, and that words $\overleftarrow{L}_j(U')$ and $R_j(U')$ are positive, we deduce inequalities

$$\text{diff}(U') = |U'_a| \geq |\overleftarrow{L}_j(U')| = |\overleftarrow{L}(U \circ h[t]h'')|$$
$$\geq |\overleftarrow{L}(U \circ h[t])| + |h''| > s(t)c + |U_a| + |h''|. \tag{7.5}$$

On the other hand we have the inequality (7.3) as in Case 1.1, a contradiction with (7.5).

Case 2. The case when $z = R_j$ is completely similar to Case 1.

Case 3. Let $z = L_j$. As in Case 1, we assume t being minimal. Then the t-th rule in h belongs to $\mathcal{S}(2) \cup \mathcal{S}(4) \cup \mathcal{S}(1)$.

Case 3.1. Suppose the t-th rule in h belongs to $\mathcal{S}(2)$. Let $h = h[t]h_1$. If all rules in h_1 belong to $\mathcal{S}(2)$, we prove the statement as in Case 1.

If $h_1 = h_2 \tau h_3$ where h_2 is a word over $\mathcal{S}(2)$, $\tau \in \mathcal{S}(23)$ then by Lemma 7.4 $|L_j(U \circ h[t-1])| < |L_j(U \circ h[t]h_2)|$ which is impossible since $|L_j(U \circ h[t]h_2)| = 0$ (because rules from $\mathcal{S}(23)$ lock $L_j P_j$-sectors).

Hence we can assume that $\tau \in \mathcal{S}(12)$.

Case 3.1.1. Assume that $h[t]h_2$ is not a word in $\mathcal{S}(2)$. Then it contains a rule from $\mathcal{S}(12)$ or from $\mathcal{S}(23)$. The first option is impossible by Lemma 7.7. Hence $h[t] = h[q]\tau'h'$ where $\tau' \in \mathcal{S}(23)$, h' is a word in $\mathcal{S}(2)$.

By Lemma 7.4, the value $|\overleftarrow{L}_j(U \circ h[p])|$ must decrease when p runs from q to $t + |h_2|$ because $|\overleftarrow{L}_j(U \circ h[t]h_2)| = 0$ (since τ locks $\overleftarrow{L}_j L_j$-sectors). Hence the sum $|\overleftarrow{L}_j(U \circ h[p])| + |L_j(U \circ h[p])|$ cannot increase. Therefore $|\overleftarrow{L}_j(U \circ h[q])| \geq |L_j(U \circ h[t])| > s(t)c + |U|$ since $|L_j(U \circ h[t]h_2)| = 0$ (τ' locks $L_j P_j$-sectors).

Since $s(q) \leq s(t)$, the number q satisfies the hypothesis of the lemma for $z = \overleftarrow{L}_j$ (Case 1 of the proof). But this contradicts part (a) of the lemma which we already proved in Case 1 (by this part, in Case 1, the brief history of the suffix of h that starts after $h[q]$ cannot contain (12)).

Case 3.1.2. Now assume that $h[t]h_2$ is a word in $\mathcal{S}(2)$. Hence $s(t) = 0$ (since $h[t]$ does not contain rules from $\mathcal{S}[34]$). By the assumption of the lemma

$$|\overleftarrow{L}_j(U \circ h[t])| + |L_j(U \circ h[t])| \geq |L_j(U \circ h[t])| > |U| \geq |\overleftarrow{L}_j(U)| + |L_j(U)|.$$

The strict inequality means that $|\overleftarrow{L}(U \circ h[p])|$ cannot monotonically decrease when p runs from 0 to $t+|h_2|$ (an age of (2)). Hence $|\overleftarrow{L}_j(U \circ h_t h_2| > 0$ by Lemma 7.4. But this contradicts the fact that τ (as every rule in $\mathcal{S}(12)$) locks $\overleftarrow{L}_j L_j$-sectors. This contradiction completes Case 3.1.

Case 3.2. The case when the t-th rule in h belongs to $\mathcal{S}(4)$ is similar, but we need to substitute $\mathcal{S}(4), \mathcal{S}(45), \mathcal{S}(34)$ for $\mathcal{S}(2), \mathcal{S}(12), \mathcal{S}(23)$.

Case 3.3. Suppose that the t-th rule in h belongs to $\mathcal{S}(1)$. Let $h = h[t]h'$. If all rules in h' belong to $\mathcal{S}(1)$ then we can argue as in Case 1, apply Lemma 7.4, and deduce parts (a), (b) of the conclusion of the lemma, and the inequality

$$|L_j(U \circ h[d])| \geq |L_j(U \circ h[t])| + (d-t)$$

for every $d \geq t$ which implies part (c).

Case 3.3.1. Assume that the letter (1) in the brief history of h' is followed by (12). Let $h' = h'[t']h_1$ where $h'[t']$ is the maximal prefix of h' which is a word over $\mathcal{S}(1)$. Since $|h_1| > 0$, $d \geq t+t'+1$. As above, we obtain by Lemma 7.4,

$$\begin{aligned}|L_j(U \circ h[t+t'+1])| = |L_j(U \circ h[t+t'])| = |L_j(U \circ h[t])| + t' \\ \geq |U| + s(t)c + t' + 1 \geq t' + 1 > 0,\end{aligned} \quad (7.6)$$

and all three parts of the statement of the lemma are true if $d = t+t'+1$. So suppose that $t+t'+1 < d$. Then

$$\text{diff}(U \circ h[t+t'+1]) > |U_a| + s(t)c + t' - |P_j(U \circ h[t+t'+1])|,$$

but $\text{diff}(U \circ h[t+t'+1]) \leq |U_a| + s(t)c$ by the same argument as in Case 1. Therefore

$$|P_j(U \circ h[t+t'+1])| \geq t'+1. \quad (7.7)$$

Then, by Lemma 7.7, the brief history of h_1 starts with (12)(2) (it cannot start with (12)(23) since $|L_j(U \circ h[t+t'+1])| > 0$ by (7.6)). Let $h_1 = h_1[t_1]h_2$ where $h_1[t_1]$ is the maximal prefix of h_1 with brief history (12)(2).

Case 3.3.1.1. Suppose that h_2 is empty. Then part (a) of the lemma is obviously true. Part (b) is also true because $|\overleftarrow{L}_j(h[q])| + |L_j(h[q]|$ stays the same for all $q \geq t+t'+1$ (Lemma 7.9 (1)), and $|L_j(U \circ h[t+t'+1])| > 0$ by (7.6).

In order to prove part (c), notice that by Lemma 7.9 (2) the sequences $|L_j(U \circ h[p])|$, $p = t+t'+1, t+t'+2...$, is monotonically decreasing. Therefore $d - t - t' \leq |L_j(U \circ h[t+t'+1])|$. Therefore by Lemma 7.9 (1) and by (7.7)

$$\begin{aligned}|U \circ h[d]| &= |U \circ h[t+t'+1]| \\ &> |L_j(U \circ h[t+t'+1])| + |P_j(U \circ h[t+t'+1])| \\ &\geq (d-t-t') + (t'+1) \geq d-t+1\end{aligned} \quad (7.8)$$

which proves part (c).

Case 3.3.1.2. Now suppose that h_2 is not empty, and $d > t + t' + t_1$. If $|h_2| = 1$ then $h_2 \in \mathcal{S}(23)$ and parts (a), (b) follow from Case 3.3.1.1, part (c) follows from (7.8). So suppose that $|h_2| > 1$, $h_2 = h_2[t_2]h_3$ where $h_2[t_2]$ is the maximal prefix of h_2 with brief history $(23)(3)$.

Since the brief history of h_2 starts with (23) and rules of $\mathcal{S}(23)$ lock $L_j P_j$-sectors, $|L_j(U \circ h[t + t' + 1 + t_1])| = 0$. As in Case 3.3.1.1 we have that $|\overleftarrow{L}_j(U \circ h[t + t' + 1 + t_1])| \neq 0$. Since rules from $\mathcal{S}(3)$ are not active for $\overleftarrow{L}_j$-sectors, the length $|\overleftarrow{L}_j(U \circ h[q])|$ stays the same (and > 0) while q runs from $t + t' + 1 + t_1$ to $t + t' + 1 + t_1 + t_2$.

Suppose that h_3 is not empty, then it starts with a rule in $\mathcal{S}(34)$. Since rules in $\mathcal{S}(34)$ lock $P_j R_j$-sectors, $P_j(U \circ h[t + t' + 1 + t_1 + t_2])$ is empty.

Therefore we obtain by Lemma 7.9

$$\begin{aligned}
s(t)c + |U| \\
&< |\overleftarrow{L}_j(U \circ h[t])| + |L_j(U \circ h[t])| \\
&\leq |\overleftarrow{L}_j(U \circ h[t + t' + 1])| + |L_j(U \circ h[t + t' + 1])| \\
&= |\overleftarrow{L}_j(U \circ h[t + t' + 1 + t_1 + t_2])| + |L_j(U \circ h[t + t' + 1 + t_1 + t_2])| \\
&\leq \mathrm{diff}(U \circ h[t + t' + 1 + t_1 + t_2]).
\end{aligned}$$

But, as in Case 1, $\mathrm{diff}(U \circ h[t + t' + 1 + t_1 + t_2]) \leq \mathrm{diff}(U) + s(t)c$, a contradiction.

Thus h_3 is empty. Parts (a) and (b) of the lemma follow as in Case 3.3.1.1. Also notice that since rules in $\mathcal{S}(3)$ lock $L_j P_j$-sectors, we have

$$|U \circ h[d]| \geq |\overleftarrow{L}_j(U \circ h[d])| + |P_j(U \circ h[d])|. \tag{7.9}$$

Since rules in $\mathcal{S}(3)$ are not active for $\overleftarrow{L}_j L_j$-sectors,

$$|\overleftarrow{L}_j(V)| = |\overleftarrow{L}_j(U \circ h[t + t' + 1 + t_1])|.$$

By Lemma 7.9, since rules in $\mathcal{S}(12)$ lock $\overleftarrow{L}_j L_j$-sectors and rules in $\mathcal{S}(23)$ lock $L_j P_j$-sectors, we have

$$\begin{aligned}
|\overleftarrow{L}_j(U \circ h[t + t' + 1 + t_1])| &= |L_j(U \circ h[t + t' + 1])| \\
&\geq |L_j(U \circ h[t])| + t' > |U| + s(t)c + t'.
\end{aligned}$$

Hence

$$|\overleftarrow{L}_j(U \circ h[d])| > |U| + s(t)c + t'. \tag{7.10}$$

Also notice that by Lemma 7.4,

$$|\overleftarrow{L}_j(U \circ h[d])| = t_1 - 1. \tag{7.11}$$

By Lemma 7.4, since rules of $\mathcal{S}(2) \cup \mathcal{S}(23)$ lock $R_j \overrightarrow{R}_j$-sectors, we have that

$$R_j(U \circ h[d]) = |h_2| - 1 = d - t - t' - t_1 - 1. \qquad (7.12)$$

Also notice that

$$|\overleftarrow{L}_j(U \circ h[d])| + |R_j(U \circ h[d])| - |P_j(U \circ h[d])| \le \operatorname{diff}(U \circ h[d])$$
$$= \operatorname{diff}(U \circ h[t])) \le \operatorname{diff}(U) + s(t)c.$$

Hence, by (7.10) and (7.12),

$$\begin{aligned}|P_j(U \circ h[d])| &\ge (|\overleftarrow{L}_j(U \circ h[d])| - s(t)c - \operatorname{diff}(U)) + |R_j(U \circ h[d])| \\ &\ge t' + d - t - t' - t_1 - 1 + |U| - \operatorname{diff}(U) \\ &\ge d - t - t_1 + 1.\end{aligned} \qquad (7.13)$$

Now using (7.11) and (7.13), we get

$$|U \circ h[d]| \ge |\overleftarrow{L}_j(U \circ h[d])| + |P_j(U \circ h[d])| \ge (t_1 - 1) + d - t - t_1 + 1 = d - t$$

as required.

Case 3.3.2. The remaining case is when the letter (1) in the brief history of h is followed by (51). This case can be analyzed as above, but the analysis is shorter, because the longest possible suffix of the brief history under consideration now is (1)(51)(5) (instead of (1)(12)(2)(23)(3) in Case 3.3.1). This subword cannot be followed by (45) by the same argument as in Case 3.3.1.2 (see the argument showing that the subword (1)(12)(2)(23)(3) cannot precede the letter (34)) because rules from $\mathcal{S}(45)$ lock $P_j R_j$-sectors. The lemma is proved. □

Lemma 7.11. *Let $U \bullet h = V$ be a j-standard computation of $\mathcal{S}$, $j \ne 1$. Then*

(1) if h contains a letter from $\mathcal{S}(34)$ then such a letter occurs in the prefix $h[30(|U| + c)]$.

(2) if $|h| \ge |V| + 30(|U| + c)$ then $h[30(|U| + c)]$ contains a subword starting with a rule from $\mathcal{S}(34)$ and ending with a rule from $\mathcal{S}(12) \cup \mathcal{S}(51)$.

Proof. (1) Suppose that h contains rules from $\mathcal{S}(34)$. Let $l = 30(|U| + c)$. Suppose, by contradiction, that $h[l]$ does not contain a rule from $\mathcal{S}(34)$.

If there exists $t \le l$ that satisfies the conditions of Lemma 7.10 (for some $z \in \{\overleftarrow{L}_j, L_j, R_j\}$) then $s(t) = 0$, and by Lemma 7.10 h does not contain letters from $\mathcal{S}(34)$, a contradiction.

Therefore for every $t = 1, ..., l$,

$$|z(U \circ h[t])| \le |U|$$

for every $z \in \{\overleftarrow{L}_j, L_j, R_j\}$. Then by Lemma 7.5, the length of an arbitrary subword of $h[l]$ over $\mathcal{S}(i)$, $i \in \{1, 2, 3, 4, 5\}$ is at most $2|U|$. Since (34) is not in the brief history of $h[l]$, the brief history of $h[l]$ contains at most 5 occurrences of $i \in \{1,2,3,4,5\}$ by Lemma 7.7. Thus the length l of $h[l]$ does not exceed $10|U| + 4 < 30(|U| + c)$ contrary to the definition of l.

(2) If there are two letters from $\mathcal{S}(34)$ in $h[l]$ then there is a letter from $\mathcal{S}(12)$ and a letter from $\mathcal{S}(51)$ between them by Lemma 7.7, and the claim of the lemma is true.

Therefore, we will suppose that $h[l]$ contains exactly one letter from $\mathcal{S}(34)$. By Lemma 7.7, the length of the brief history of $h[l]$ is at most 19, and the brief history contains at most 10 maximal subwords over $\mathcal{S}(1) \cup ... \cup \mathcal{S}(5)$.

Suppose that some number $t \leq |h|$ satisfies an assumption of Lemma 7.10 for some $z \leq \{\overleftarrow{L_j}, L_j, R_j\}$, and $s(t) = 1$. Take the minimal t with this property. Then $|z(U \circ h[p])| \leq c + |U|$ for $p < t$, for any $z \in \{\overleftarrow{L_j}, L_j, R_j\}$.

Then, as in the proof of part (1), $t - 1 \leq 10(2|U| + 2c) + 9 < 30(|U| + c)$ because the number of ages of $(1), (2), \ldots (5)$ is at most 10 in the history of $h[t-1]$. This inequality together with the inequality $|U \circ h| \geq |h| - t$ given by Lemma 7.10 (part (c)) implies that $|h| \leq |U \circ h| + 30(|U| + c)$ contrary to the assumption of part (2).

Hence there is no number t between 1 and $|h|$ such that $|z(U \circ h[t])| \geq c + |U|$ for some $z \in \{\overleftarrow{L_j}, L_j, R_j\}$. Then, as in (1), Lemmas 7.5 and 7.7 provide us with the inequality $|h| \leq 10(2|U| + 2c) + 9 < 30(|U| + c)$ contrary to the assumption of the lemma. The lemma is proved. □

Lemma 7.12. *Let $U \bullet h = V$ be a j-standard computation of $\mathcal{S}$ and $j \neq 1$ or a j-standard computation of $\bar{\mathcal{S}}$ (for any j).*

Assume that h contains an occurrence of a rule τ_1 from $\mathcal{S}(34) \cup \bar{\mathcal{S}}(34)$ preceding an occurrence of a rule τ_2 from $\mathcal{S}(12) \cup \mathcal{S}(51) \cup \bar{\mathcal{S}}(12) \cup \bar{\mathcal{S}}(51)$: $h = h_1 \tau_1 h_2 \tau_2 h_3$. Then the words $z(U \circ h_1 \tau_1 h_2)$, $z \in \{L_j, P_j\}$, and $z(U \circ h_1)$, $z \in \{\overleftarrow{L_j}, R_j\}$ are completely determined by the word $\tau_1 h_2$.

Proof. If $j = 1$ and h is a word over $\bar{\mathcal{S}}$, then all the words listed in the formulation of the lemma are empty, so the statement is trivially true.

Suppose that $j \neq 1$. We may suppose that h_2 does not contain rules from $\mathcal{S}(34) \cup \mathcal{S}(12) \cup \mathcal{S}(51) \cup \bar{\mathcal{S}}(34) \cup \bar{\mathcal{S}}(12) \cup \bar{\mathcal{S}}(51)$. By Lemma 7.7, the brief history of $\tau_1 h_2 \tau_2$ is either $(34)(3)(23)(2)(12)$ or $(34)(4)(45)(5)(51)$. These cases are similar, so let the brief history be equal to $(34)(3)(23)(2)(12)$.

Let $h_2 = h' \tau h''$ where $\tau \in \mathcal{S}(23)$ (or $\tau \in \bar{\mathcal{S}}(23)$). Since rules from $\mathcal{S}(23)$ lock $L_j P_j$-sectors, and each rule $\tau(2, r, i)^{\pm 1}$ of $\mathcal{S}(2)$ (of $\bar{\mathcal{S}}(2)$) multiplies inner parts of the $L_j P_j$-sectors of admissible words by the letter $a_i(L_j)^{\mp 1}$ (resp. by $\bar{a}_i(L_j)^{\mp 1}$) on the left (Lemma 2.6), the word $L_j(U \circ h' \tau h'')$ is a copy of h'' written from right to left. Therefore the word $L_j(U \circ h_1 \tau h_2)$ is completely determined by h_2. Similarly, the word $P_j(U \circ h_1 \tau h_2)$ is completely determined by h' (and, in turn, by h_2) because rules from $\mathcal{S}(34)$ lock $P_j R_j$-sectors, rules from $\mathcal{S}(3)$ are active with respect to $P_j R_j$-sectors, and rules from $\mathcal{S}(23) \cup \mathcal{S}(2)$ are not active with respect to these sectors.

Similarly, since τ_2 locks $\overleftarrow{L}_j$-sectors and R_j-sectors (as do all rules from $\mathcal{S}(12)$), rules from $\mathcal{S}(2)$ are active with respect to these sectors, and rules from $\mathcal{S}(3) \cup \mathcal{S}(23)$ are not active with respect to these sectors, the word $\overleftarrow{L}_j(U \circ h_1)$ (resp. $R_j(U \circ h_1)$) is completely determined by $\tau_1 h_2$. $\square$

Lemma 7.13. *Let $U \bullet h = V$ be a j-standard computation of $\mathcal{S}$ and $j \neq 1$. Suppose that $|h| \geq |V| + 30(|U|+c)$. Then $h[30(|U|+c)]$ can be decomposed as $h[30(|U|+c)] = h_1 \tau_1 h_2 \tau_2 h_3$ where $\tau_1 \in \mathcal{S}(34)$, $\tau_2 \in \mathcal{S}(12) \cup \mathcal{S}(51)$, h_2 contains no letters from $\mathcal{S}(34) \cup \mathcal{S}(12) \cup \mathcal{S}(51)$. Moreover the word $P_j(U \circ h_1 \tau_1 h_2 \tau_2)$ is positive for any such decomposition of $h[30(|U|+c)]$.*

Proof. By Lemma 7.11 there exists a decomposition

$$h[30(|U|+c)] = h_1 \tau_1 h_2 \tau_2 h_3$$

where $\tau_1 \in \mathcal{S}(34)$, $\tau_2 \in \mathcal{S}(12) \cup \mathcal{S}(51)$. Taking h_2 of minimal length, we can assume that h_2 contains no letters from $\mathcal{S}(34) \cup \mathcal{S}(12) \cup \mathcal{S}(51) \cup \mathcal{S}(1)$. So we only need to show that $P_j(U \circ h_1 \tau_1 h_2 \tau_2)$ is positive. Consider the homomorphism γ from the free group generated by $\mathcal{A} \cup \mathcal{K}$ into the free group generated by $\{a_1, ..., a_{\bar{m}}\}$ which maps $a_i(z)$ into a_i for every z, and maps all other letters to 1. It is clear that for every rule $\tau \in \mathcal{S} \setminus \mathcal{S}(34)$, and every admissible word W in the domain of τ, $\gamma(W \circ \tau) = \gamma(W)$. Let U' be the subword of U stating with a P_j-letter and ending with a $\overrightarrow{R}_j$-letter. Then $U' \circ h = V'$ where V' is the corresponding subword in V. Moreover for every t between 1 and $|h|$, and $z \in \{P_j, R_j\}$, the zz_+-sector in $U' \circ h[t]$ coincides with the zz_+-sector in $U \circ h[t]$. Since rules in $\mathcal{S}(34)$ lock $P_j R_j$-sectors,

$$\gamma(U' \circ h_1 \tau_1) = \gamma(R_j(U' \circ h_1 \tau_1)).$$

The word in the right hand side of this equality is a copy of $R_j(U' \circ h_1 \tau_1)$ and is positive by definition of admissible words. Hence $\gamma(U' \circ h_1 \tau_1 h_2) = \gamma(R_j(U' \circ h_1 \tau_1))$ (we use here that h does not contain rules from $\mathcal{S}(1)$) is a positive word. But the rules in $\mathcal{S}(12) \cup \mathcal{S}(51)$ lock $R_j \overrightarrow{R}_j$-sectors, so $\gamma(U' \circ h_1 \tau_1 h_2)$ is a copy of $P_j(U' \circ h_1 \tau_1 h_2) = P_j(U \circ h_1 \tau_1 h_2)$. Hence $P_j(U \circ h_1 \tau_1 h_2)$ is a positive word as required. $\square$

To formulate the next lemma, we introduce a natural homomorphism λ from the free group F_0 with basis $\mathcal{A} \cup \bar{\mathcal{A}}$ onto the factor group H_0 of F_0 over the normal closure of all $\mathcal{G}$-relations. Thus H_0 is the free product of a copy of $\mathcal{G}$ generated by $a_1(P_1), ..., a_m(P_1)$ and a free group generated by all other a-letters.

Lemma 7.14. *Let $U \bullet h = V$ be a 1-standard $\mathcal{G}$-computation of $\mathcal{S}$. Suppose that $h = h_1 \tau_1 h_2 \tau_2 h_3$ where $\tau_1 \in \mathcal{S}(34)$, $\tau_2 \in \mathcal{S}(12) \cup \mathcal{S}(51)$. Assume that h satisfy the conclusion of Lemma 7.7. Then the words $z(U \circ h_1 \tau_1)$, $z \in \{\overleftarrow{L}_1, R_1\}$, $L_1(U \circ h_1 \tau_1 h_2)$ and the λ-image of $P_1(U \circ h_1 \tau_1 h_2)$ are determined by the word $\tau_1 h_2 \tau_2$.*

Proof. The proof of the assertion about $z(U \circ h_1\tau_1)$, $z \in \{\overleftarrow{L}_1, R_1\}$ and $L_1(U \circ h_1\tau_1 h_2)$ coincides with the corresponding proof in Lemma 7.12. Let U' be the P_1R_1-sector in U. Then $P_1(U \circ h[t]) = P_1(U' \circ h[t])$ for every t. Notice also that $P_1(U' \circ h_1)$ is equal to 1 modulo $\mathcal{G}$ because rules from $\mathcal{S}(34)$ lock the P_1R_1-sectors. Hence, modulo $\mathcal{G}$, $U' \circ h_1$ is a 2-letter word over $\mathcal{K}$. Therefore $U' \circ h_1\tau_1 h_2 = (U' \circ h_1\tau_1) \circ h_2$ is determined by h_2 modulo $\mathcal{G}$ as desired. Hence the λ-image of $P_1(U \circ h_1\tau_1)$ is determined by the word $\tau_1 h_2 \tau_2$. $\square$

7.3 Tame and wild computations

Recall that for every positive word w in $\{a_1, ..., a_{\bar{m}}\}$ we have defined a word $\Sigma(w)$ in (2.3) which has the base $\tilde{\Sigma}$ defined in (2.2).

Let U be an admissible word for $\mathcal{S}$ with a base $\tilde{\Sigma}^s K_1$. Then the word U and any ring computation $U \bullet h = V$ over $\mathcal{S} \cup \bar{\mathcal{S}}$ will be called *tame*. All non-tame admissible words U whose bases start and end with the same letter will be called *wild*. If U is a wild admissible word then any ring computation $U \bullet h = V$ will be called *wild* as well.

Recall that every admissible word for $\mathcal{S} \cup \bar{\mathcal{S}}$ has an $\bar{\mathcal{E}}$-coordinate and an Ω-coordinate. All k-letters in an admissible word U have the same coordinates r, i, so we can write $U = U'(r,i)$ where U' is the word U with $(\bar{\mathcal{E}}, \Omega)$-coordinates removed from all k-letters.

Lemma 7.15. *Let $U \bullet h = V$ be a tame computation of $\mathcal{S}$, with the base $U = \tilde{\Sigma}^s K_1$. Suppose that $|h| \geq |V| + 30(|U| + c)$. Then the prefix $h[30(|U| + c)]$ has the form $h_1 \tau h_2$ where $\tau \in \mathcal{S}(12) \cup \mathcal{S}(51)$ and $U \circ h_1$ or $U \circ h_1 \tau$ has the form $\Sigma(u, v)^s K_1$ where u and v are positive words over $\{a_1, ..., a_{\bar{m}}\}$, and $\Sigma(u, v)$ is an admissible word of the following form*

$$(\prod_{j=1}^{sN/2} K_{2j-1} L_{2j-1} v_{2j-1} P_{2j-1} u_{2j-1} R_{2j-1} K_{2j}^{-1} R_{2j}^{-1} u_{2j}^{-1} P_{2j}^{-1} v_{2j}^{-1} L_{2j}^{-1})(\emptyset, 1) \quad (7.14)$$

where v_k, $u_{k'}$ are copies of v and u under the substitutions $a_i \mapsto a_i(L_k)$, $a_i \mapsto a_i(P_{k'})$ for every k and for $k' \neq 1 (\bmod N)$; u_{1+Nl} is equal to a copy of u modulo $\mathcal{G}$.

Proof. Let U' be any admissible subword of U with base $\overleftarrow{L}_j L_j P_j R_j \overrightarrow{R}_j$. Let V' be the corresponding subword in V. Then $U' \bullet h = V'$ is a j-standard computation, and we can apply Lemma 7.13. This gives us the desired occurrence of $\tau \in \mathcal{S}(12) \cup \mathcal{S}(51)$ in the prefix $h[30(|U|+c)]$. Then, depending on whether $\tau \in \mathcal{S}^-(12) \cup \mathcal{S}^+(51)$ or $\tau \in \mathcal{S}^+(12) \cup \mathcal{S}^-(51)$, either $U \circ h_1 \tau$ or $U \circ h_1$ has the form (7.14). The fact that all words u_i are copies of each other and all words v_i are copies of each other follows from Lemmas 7.12 and 7.14 (in case $j = 1$) since the histories of the j-standard computations are all equal to h (notice that Lemma 7.14 can be applied because h satisfies the conclusion of Lemma 7.7 since there exists a j-standard computation $U' \bullet h = V'$ for $j \neq 1$). The fact that u_i and v_i are positive words follows from Lemma 7.13 and from the fact that words $U \circ h[t]$ are admissible for $\mathcal{S}$. □

Lemma 7.16. *Let $|\mathcal{S}|$ be the number of rules in $\mathcal{S}$. Let $U \bullet h = V$ be a wild ring computation of $\mathcal{S}$. Suppose that $|h| \geq |\mathcal{S}|+1$, h contains a rule from $\mathcal{S}(34)$. Then $U^{\pm 1}$ contains an admissible subword with base $\overleftarrow{L}_j L_j P_j R_j \overrightarrow{R}_j$ for some $j \geq 1$. If in addition the brief history of h contains one of the letters $(1), (12), (2), (23), (45), (5), (51)$, then U contains such an admissible subword with base $\overleftarrow{L}_j L_j P_j R_j \overrightarrow{R}_j$ for some $j > 1$.*

Proof. If the base of U does not contain L-, P- and R-letters, then the words $U \circ h[t]$, $t = 1, ..., |h|$ differ only by the $(\bar{\mathcal{E}}, \Omega)$-coordinates (since rules of $\mathcal{S}$ do not change $K_j K_j^{-1}$- and $K_j^{-1} K_j$-sectors of admissible words). Since the computation $U \bullet h = V$ is a ring computation, the labels of the annuli of the ring, which correspond to words $U \circ h[t]$, $U \circ h[t']$ ($t \neq t'$) can not be equal modulo $\mathcal{G}$-relations. Then Lemma 6.1 shows that the number of the annuli cannot be greater than the number of rules of $\mathcal{S}$. Hence $|h|$ does not exceed $|\mathcal{S}|+1$.

Assume now that the base contains L_j-, P_j- or R_j-letter for some j. Then since rules from $\mathcal{S}(34)$ lock $L_j P_j$ and $P_j R_j$-sectors, the base of $U^{\pm 1}$ must contain a subword $L_j P_j R_j$. By Lemma 2.2 this subword must be inside a subword of the form $\overleftarrow{L}_j L_j P_j R_j \overrightarrow{R}_j$ and the first claim of the lemma true.

Suppose that $j = 1$. Notice that $\overrightarrow{R}_1 = \overrightarrow{R}_2^{-1}$. Now if the brief history of h contains letter from $\{(1), (12), (2), (23), (51)\}$ then the corresponding rules in h lock $R_2 \overrightarrow{R}_2$-sectors. Therefore the base of $U^{\pm 1}$ contains R_2^{-1} which implies, as before, that it contains a subword of the form $\overleftarrow{L}_2 L_2 P_2 R_2 \overrightarrow{R}_2$, as required.

Similarly notice that $\overleftarrow{L}_1 = \overleftarrow{L}_N$, and rules from $\mathcal{S}(45) \cup \mathcal{S}(5)$ lock $\overleftarrow{L}_j L_j$-sectors for every j. So if the brief history of h contains (45) or (5) then the base of $U^{\pm 1}$ contains the subword $\overleftarrow{L}_N L_N P_N R_N \overrightarrow{R}_N$. □

Lemma 7.17. *Let $U \bullet h = V$ be a wild or tame ring computation of $\mathcal{S}$, $|h|$ exceeds $|\mathcal{S}| + 1$. Suppose that h contains a rule from $\mathcal{S}(34)$ and a rule from $\mathcal{S}(1) \cup \mathcal{S}(12) \cup \mathcal{S}(51)$. Then the computation is tame.*

Proof. The presence of a rule locking $\overleftarrow{L}_j L_j$-sector (locking $R_j \overrightarrow{R}_j$-sector), implies that the letter $\overleftarrow{L}_j$ (letter $\overrightarrow{R}_j$) can occur in the base of U only in subwords $(\overleftarrow{L}_j L_j)^{\pm 1}$ (resp. in subwords $(R_j \overrightarrow{R}_j)^{\pm 1}$). Together with the argument from the proof of Lemma 7.16, this implies that arbitrary letter of the base of U has a uniquely determined right neighbor, which coincides with the right neighbor of this letter in the base of a tame ring. □

Lemma 7.18. *Let $U \bullet h = V$ be a wild ring computation of $\mathcal{S}$, h contains a rule from $\mathcal{S}(34)$. Then $|h|$ is bounded by a linear function in $|U| + |V|$.*

Proof. By Lemma 7.17, we may assume that the brief history br(h) of h does not contain the letters (12) and (51). Suppose in the beginning, that br(h) has at least one of the letters (1), (2), (23), (45), (5). Then, applying Lemmas 7.16 and 7.11, we conclude that $|h|$ is bounded by a linear function in $|U| + |V|$.

So, we may assume that br(h) contains no letters from the list

$$\{(1), (12), (2), (23), (45), (5), (51)\},$$

i.e. it is a word in letters (3), (34), (4). By Lemma 7.8 it has no subwords (34)(4)(34). Thus, we may suppose that br(h) is a subword of the word (4)(34)(3)(34)(4). Hence

$$h = h_{(4)} h_{(34)} h_{(3)} h'_{(34)} h'_{(4)}$$

where h_α is a word in $\mathcal{S}(\alpha)$ (the first and the last few subwords in this decomposition of h may be empty), $|h_{(34)}|, |h'_{(34)}| \leq 1$.

By Lemma 7.16, we may assume that $U^{\pm 1}$ contains an admissible subword U' with the base $\overleftarrow{L_j} L_j P_j R_j \overrightarrow{R_j}$ for some j. Consider the computation $U' \bullet h = V'$. By Lemma 7.5 applied to the $L_j P_j$-sector U'' of U', $|h_{(4)}|$ is bounded by $|U| + |V|$ since $L_j(U'' \circ h_{(4)})$ is empty.

The same is true for $|h'_{(4)}|$. Hence $|U \circ h_{(4)}|$ and $|U \circ h_{(4)} h_{(34)} h_{(3)}|$ are bounded by a linear function in $|U| + |V|$ (the application of every rule can increase the length of an admissible word with a given base only by a constant). Now again by Lemma 7.5 $|h_{(3)}|$ is bounded by a linear function in $|U| + |V|$. □

Lemma 7.19. *Let $U \bullet h = V$ be a wild computation of $\mathcal{S}$. Suppose that h does not contain rules from $\mathcal{S}(34)$. Then $|h|$ is bounded by a linear function in $|U| + |V|$.*

Proof. Lemma 7.8 (and the absence of (34)-letters in the brief history B of h) shows that br(h) does not have any of the six prohibited words mentioned in Lemma 7.8 and also (34)(3)(34). Therefore one can apply the argument of Lemma 7.7, and conclude that br(h) must be obtained by removing some letters in one of the following four words:

(i) (4)(45)(5)(45)(4),

(ii) (3)(23)(2)(12)(1)(12)(2)(23)(3),

(iii) ((3)(23)(2)(12)(1)(51)(5)(45)(4))$^{\pm 1}$,

(iv) (4)(45)(5)(51)(1)(51)(5)(45)(4)

Case (i). In this case, as above, we have the following decomposition:

$$h = h_{(4)}h_{(45)}h_{(5)}h'_{(45)}h'_{(4)}$$

where some of the factors can be empty.

Suppose that the base of U does not contain L-letters. Since rules in $S(4)$ are not active with respect to zz'-sectors where $z, z' \neq L_j$, every rule in $S(4)$ fixes U. Therefore the words $h_{(4)}$ and $h'_{(4)}$ are empty (otherwise we get a contradiction with the assumption that the computation $U \bullet h = V$ is a ring computation (rings are incompressible), so all words $U \circ h[t]$, $t = 0, 1, .., |h|$ are supposed to be different).

If the base of U contains L_j then it contains $\overleftarrow{L_j}L_j$ by Lemma 2.2. By Lemma 7.5, this gives a linear upper bound for the lengths of $h_{(4)}$ and $h'_{(4)}$ in terms of the lengths of U and V by the locking argument for $S(34)$.

If the base of U contains R_j then it contains a subword $R_j\overrightarrow{R_j}$ and the length of $h_{(5)}$ is also linearly bounded from above by Lemma 7.5. Otherwise $|h_{(5)}| = 0$ because the computation is absolutely reduced. Thus all five subwords in the decomposition of h have linearly bounded length.

Case (ii). By Lemma 7.3, we can turn $U \bullet h = V \pmod{\mathcal{G}}$ into a free computation $U \bullet h = V'$ where $V' = V \pmod{\mathcal{G}}$. All sectors in V' but the P_1R_1-sector are freely equal to the corresponding sectors in V. Using Lemma 7.5 to the computations $z(U) \bullet h = z(V')$, $z \in \{\overleftarrow{L_j}, L_j, R_j\}$, we linearly bound the lengths of maximal subwords of h over $S(i)$ ($i = 1, 2, 3$) in terms of $|z(U)| + |z(V')| \leq |U| + |V|$. For example, if $i = 1$, $z = L_j$, etc.

Cases (iii),(iv) are similar to cases (1) and (ii). □

Lemma 7.20. *Let $U \bullet h = V$ be a wild ring computation of S. Then $|h|$ is bounded by a linear function in $|U| + |V|$.*

Proof. The statement follows from Lemmas 7.18 and 7.19. □

We call an admissible word U for $\bar{S}$ (resp. S) *accepted* by $\bar{S}$ (by S) if there exists a ring computation of the form $U \circ h = \Sigma^s(w)K_1(\emptyset, 1)$, $s \neq 0$, of $\bar{S}$ (resp. of S mod $\mathcal{G}$) for some (for some positive) word w.

Lemma 7.21. *(1)If a tame admissible word U whose base starts and ends with the same letter is not accepted by S then the length of the brief history of any ring computation $U \bullet h = V$ of S is at most $|S| + 20$.*

(2) If U is accepted by S, then the length d of any ring computation

$$U \circ h = \Sigma^s(w)K_1(\emptyset, 1),$$

such that $U \circ h[d'] \neq \Sigma^s(w')K_1(\emptyset, 1)$ for a word w' and $d' < d$, is recursively bounded in terms of $|U|$.

Proof. (1) Suppose that the word U is not accepted but the length of the brief history of some ring computation $U \bullet h$ is $> |S| + 20$. Then by Lemma 7.7 $\mathrm{br}(h)$ contains an occurrence of (34) and one of the letters (1), (12), (2), (23), (45), (5), (51). Since the computation is tame, the base of U is $\tilde{\Sigma}^s K_1$ for some $s \neq 0$. Then by Lemma 7.7, the $\mathrm{br}(h)$ contains two occurrences of (34), and an occurrence of (12) or (51) between them: $h = h_1 \tau_1 h_2 \tau_2 h_3$ where $\tau_1 \in \mathcal{S}(34)$, $\tau_2 \in \mathcal{S}(12) \cup \mathcal{S}(51)$. Then by Lemmas 7.12 and Lemma 7.14, we conclude that all sectors of $U \circ h_1 \tau_1 h_2 \tau_2$ are determined by $\tau_1 h_2$, so they are copies of each other (or, in the case of $P_1 z$-sectors, copies modulo $\mathcal{G}$). Therefore, depending on whether $\tau_2 \in \mathcal{S}^+(12)$ or $\tau_2 \in \mathcal{S}^-(12)$, the word $U \circ h_1 \tau_1 h_2$ or $U \circ h_1 \tau_1 h_2 \tau_2$ has the form $\Sigma_{\emptyset,1}(\emptyset, u, v, \emptyset) K_1(\emptyset, 1)$ (see Definition 3.1) where u and v are positive words by Lemma 7.13. Let us denote this word by U'.

By Lemma 3.5, part a) (for $i = 1$), there exists a ring computation

$$U' \bullet h' = \Sigma_{\emptyset,1}(\emptyset, \emptyset, uv, \emptyset)^s K_1(\emptyset, 1) = \Sigma(uv)^s K_1(\emptyset, 1)$$

of $\mathcal{S}(1)$ and, by Lemmas 3.4 and 6.1, the corresponding ring can be united with the ring which corresponds to $U \bullet h_1 \tau_1 h_2$ or $U \bullet h_1 \tau_1 h_2 \tau_2$. Now by Lemma 7.2, the computation $U \bullet h_1 \tau_1 h_2 h'$ or $U \bullet h_1 \tau_1 h_2 \tau_2 h'$ (or another computation if the corresponding annular diagram is compressible) is a ring computation accepting U, a contradiction.

(2) First we need to check if U is tame. If it is not tame, it cannot be accepted. So assume that U is tame. By the proof of part (1) (and the definitions of compressible and reduced diagrams), in order to check if a tame word U is accepted, one needs to check only computations $U \bullet h$ whose brief history does not contain two occurrences of (34) and an occurrence of (12) or (51) between them. By Lemma 7.7, $|\mathrm{br}(h)| \leq 19$. If for some $z \in \{\overleftarrow{L}_j, L_j, R_j\}$, $j \neq 1$, and some $t \leq |h|$, the length $|z(U \circ h[t])|$ exceeds $c + |U|$ then by Lemma 7.10, part (b), one of the words $\overleftarrow{L}_j(U \circ h[d])$, $L_j(U \circ h[d])$ or $R_j(U \circ h[d])$ is not empty for every $d \geq t$. Since the words $\overleftarrow{L}_j(\Sigma^s(w) K(\emptyset, 1))$, $L_j(\Sigma^s(w) K(\emptyset, 1))$ or $R_j(\Sigma^s(w) K(\emptyset, 1))$ are empty, we can consider only computations where $|z(U \circ h[t])| \leq c + |U|$. By Lemma 7.5, the length of every subword of h over $\mathcal{S}(i)$, $i \in \{1, 2, 3, 4, 5\}$ is bounded by $2|U| + c$. This gives a bound on the length of h. $\square$

Lemma 7.22. *There exists an algorithm to check, given two admissible words U, V for $\mathcal{S}$, if there exists a ring computation $U \bullet h = V$ of $\mathcal{S}$. The length of an arbitrary such ring computation is recursively bounded in terms of $|U| + |V|$.*

Proof. By Lemmas 7.20 and 7.1, we have a linear bound for any wild ring computation of $\mathcal{S}$ which can connect U and V.

Let us assume that there is a tame ring computation connecting U and V. By Lemmas 7.15 and 7.1, we can assume that U and V have the form

(7.14) for some $s \neq 0$. Let us use the notation from Lemma 7.15. Since the word problem in $\mathcal{G}$ is decidable, u_{1+Nl}, $l = 1,...,s$, is a copy of u modulo $\mathcal{G}$, we can assume that u_{1+Nl} is a copy of u in the free group. (Clearly if $U' = U(\mod \mathcal{G})$, then a computation $U \bullet h = V$ exists if and only if a computation $U' \bullet h = V$ exists, and the corresponding rings differ only in a $\mathcal{G}$-cell attached to the boundary of one of them.) Furthermore we may suppose that the words U and V are accepted by $\mathcal{S}$ (see the proof of Lemma 7.21), so we can assume by Lemma 7.21, part (2), that $U \equiv \Sigma(w)^s K_1(\emptyset, 1)$, $V \equiv \Sigma(w')^s K_1(\emptyset, 1)$.

Notice that for every rule $\tau \in \mathcal{S}$ and every admissible word U' with the base $W_j = \overleftarrow{L_j} L_j P_j R_j \overrightarrow{R_j}$ which is in the domain of τ, $\delta(U') = \delta(U' \circ \tau)$ for the homomorphism δ introduced before Lemma 3.1. By Lemma 3.9, we have similar equality for a ring computation (which compute modulo $\mathcal{G}$) of arbitrary length. Therefore if $\delta(w') \neq \delta(w)$, there could be no computation of $\mathcal{S}$ connecting U and V.

Assume that $\delta(w') = \delta(w)$. Then a ring computation $U \bullet h = V$ of $\mathcal{S}$ does exist by Lemmas 2.1 and 3.6. Since the equality $\delta(w) = \delta(w')$ is effectively verifiable (the word problem in $\bar{\mathcal{G}}$ is decidable), we can effectively verify (using Lemma 7.1) if there exists a ring computation of $\mathcal{S}$ connecting U and V. By Lemma 6.1, all rings of the first type corresponding to computations $U \bullet h' = V$ for some h', have boundary labels U and V. (These boundary labels have no x-letters by Lemma 6.1 since $U \equiv \Sigma(w)^s K_1(\emptyset, 1)$, $V \equiv \Sigma(w')^s K_1(\emptyset, 1)$.) It follows from the definition of compressible and reduced diagrams, that the heights of all of these rings must be equal. This height is recursively bounded. Indeed, the above construction recursively bounds the number of $\mathcal{S}(34)$-rules in any ring computation connecting U and V, because there are no (θ, k)-cells which are higher than those corresponding to rules from $\mathcal{S}(34)$. Then arguing as in the proof of Lemma 7.21, we get a recursive upper bound for the length of the history. The lemma is proved. $\square$

7.4 Computations of $\bar{\mathcal{S}}$

Lemma 7.23. *For any $s \geq 1$, and any two admissible words U, V of $\bar{\mathcal{S}}$ there is an algorithm to decide whether there exists a computation $U \bullet h = V$ of $\bar{\mathcal{S}}$, and $br(h)$ is of length $\leq s$. The lengths of all ring computations of this form are recursively bounded by a function of $|U| + |V|$.*

Proof. If U and V have different bases then there is no computation connecting U and V. So we can assume that U and V have the same bases.

We can assume that h is reduced and write it as $h \equiv h_1...h_s$ where h_i are words over $\bar{\mathcal{S}}(\omega_i)$, $\omega_i \in \{1, 12, ...51\}$, $i = 1,...,s$. There are only finitely many choices for the sequence $\omega_1,...,\omega_s$. So we can fix one of them. We can also assume that if $\mathcal{S}(\omega_i)$ consists of transition rules then h_i is of length 1, so there are finitely many choices for such h_i. Fix one of these choices.

By Lemma 2.7, for each of the other h_i, $i = 1,...,s$, there exist two sequences of numbers $\{\varepsilon_{i,z}, \delta_{i,z} \in \{-1,0,1\} \mid z \in \tilde{\mathcal{K}}\}$, such that for every admissible word W of the form $V(g,r)$, and any group word w_i in $\{a_1,...,a_{\bar{m}}\}$ there exists a word h_i in $\bar{S}(g)$ satisfying the following properties:

1. h_i is a copy of w_i;

2. W is in the domain of h_i;

3. $W \circ h_i$ is obtained from W by all replacements of the form

$$\bar{z}(r,g) \to w_i(z_-)^{\varepsilon_{i,z}} \bar{z}(r,g) w_i(z)^{\delta_{i,z}},$$

where z runs over all letters in the base of W, $\varepsilon_{i,z} \in \{-1,1\}$.

Thus, by Lemmas 2.6 and 2.7, applying h to U results in multiplying the interior of each sector of U on the left and the right by products of copies of unknown words $w_i(z)$, their inverses, and constant words corresponding to the transition rules. The result of this multiplication will be the interior of the corresponding sector in V. Hence the existence of h such that $U \circ h = V$ is equivalent to the existence of a solution of one of a finite systems of equations in the free group (each system corresponds to the choice of the brief history of h and the choice of transition rules in h; the equations correspond to the ages in the brief history and the unknowns are the words $w_i(z)$).

It remains to apply a theorem of Makanin [Mak] saying that there exists an algorithm checking if a system of equations over a free group has a solution. The heights of all ring computations of this form must be equal by the definition of compressible and reduced diagrams. This height is recursively bounded by Lemma 6.16, because the above construction gives the list of computations (if any exist) with arbitrary distribution of at most s transition rules; and this allows us to effectively select those whose rings are incompressible. The lemma is proved. $\square$

Lemma 7.24. *(1) If a tame word U is not accepted by $\bar{S}$ then the length of the brief history of any ring computation $U \bullet h = V$ of $\bar{S}$ is at less than 19 with at most one occurrence of (34).*

(2) There is an algorithm which checks whether a given admissible word is accepted by $\bar{S}$, this algorithm also finds an accepting computation if it exists.

Proof. (1) Let the base of U be $\tilde{\Sigma}^l K_1$ for some $l \neq 0$. Then using Lemma 7.7 as in the proof of Lemma 7.21, if the length of the brief history of a computation $U \bullet h$ is bigger than 19 or if the number of occurrences of (34) in the brief history is at least 2, there exists t between 1 and $|h|$ such that

$$U \circ h[t] = \bar{\Sigma}_{\emptyset,1}(\emptyset, u, v, \emptyset)^l \bar{K}_1(\emptyset, 1) = \Sigma_{\emptyset,1}(\emptyset, u, v, \emptyset)^l K_1(\emptyset, 1).$$

By Lemma 3.8, part b), there exists a computation

$$U \bullet h[t]h' = \Sigma(uv)^l K_1(\emptyset, 1).$$

By Lemma 6.16, this computation (or another computation if the corresponding annular diagram is not reduced) is a ring computation, so U is accepted. This contradiction proves the first part of the lemma.

(2) By Part 1 of the lemma and Lemma 7.7, we can restrict ourselves to computations $U \bullet h$ with $\mathrm{br}(h)$ containing at most one occurrence of (34) and $|\mathrm{br}(h)| \leq 19$.

Suppose that U is accepted, $U \circ h = \Sigma(w)^l K_1$ be the corresponding computation. As in Lemma 7.13, introduce a homomorphism γ of the free group with basis $\bar{A} \cup \bar{\mathcal{K}}$ onto free group over $\{a_1, \ldots, a_{\bar{m}}\}$, by the rule $\bar{a}_i(z) \mapsto a_i$, $z \to 1$, $z \in \bar{\mathcal{K}}$. Then for every $t = 0, \ldots, |h-1|$, $\gamma(U \circ h[t]) = \gamma(U \circ h[t+1])$ if $h[t+1]$ does not end with a rule from $\bar{S}(34)$. Otherwise

$$||\gamma(U \circ h[t])| - |\gamma(U \circ h[t+1])|| \leq cl.$$

Hence the length of $w^l = \gamma(\Sigma(w)^l K_1(\emptyset, 1))$ is bounded by $|U| + cl$ (since h contains only one rule from $\bar{S}(34)$). Hence $|w| \leq |U| + cl$. Therefore $|\Sigma(w)^l K_1(\emptyset, 1)|$ does not exceed $l(|U| + lc) + 4N$. Now Lemma 7.23 gives us an algorithm to check if an accepting ring computation exists. $\square$

Lemma 7.25. *Assume $U \bullet h$ be a wild ring computation of $\bar{S}$. Then $|br(h)| \leq 13$.*

Proof. We examine several cases taking into account Lemma 6.2. Let B be the base of U.

Case 1. Suppose that B contains letter $P_j^{\pm 1}$ for some $j \neq 1$.

Case 1.1. Suppose that B contains no letters except $P_j^{\pm 1}$. Since rules from $\bar{S}(\omega)$, $\omega \in \{(23), (3), (34), (4), (45)\}$ lock $P_j R_j$-sectors, these rule cannot occur in h. Rules from $\bar{S}(2) \cup \bar{S}(5)$ are not active for $P_j P_j^{-1}$- and $P_j^{-1} P_j$-sectors, so existence of these rules in h would contradict the assumption that we have a ring computation. Hence $\mathrm{br}(h)$ does not contain (2) and (5). This implies that $\mathrm{br}(h)$ cannot be longer than the words $(51)(1)(12)$, $(51)(1)(51)$ and $(12)(1)(12)$, $|\mathrm{br}(h)| \leq 3$.

Case 1.2. Suppose that B contains a subword yz where y (or z) is equal to $P_j^{\pm 1}$ but z (resp. y) is not $P_j^{\pm 1}$. We can assume that the exponent of $P_j^{\pm 1}$ in yz is 1 (otherwise we can switch from U to U^{-1}). Therefore by Lemma 6.2, yz is equal to either $L_j P_j$ or $P_j R_j$.

Case 1.2.1. Let $yz \equiv L_j P_j$, $j \neq 1$. Then by Lemma 6.2, since B starts and ends with the same letter, there are two possibilities: either B has a subword $L_j P_j P_j^{-1}$ or it contains $L_j P_j R_j$.

Case 1.2.1.1. Suppose there is a subword $L_j P_j P_j^{-1}$ in B for $j \neq 1$. Then there are no letters from $\{(34), (4), (45)\}$ in $\mathrm{br}(h)$ since the corresponding rules lock P_jR_j-sectors. The word B also has no subwords $(23)(2)(23)$ by Lemma 7.6. If B contains a subword $L_j^{-1}L_jP_jP_j^{-1}$, then br(h) cannot have letters (12) or (51) since the corresponding rules lock $\overleftarrow{L_j}L_j$-sectors. If B has a subword $\overleftarrow{L_j}L_jP_jP_j^{-1}$, then br($h$) cannot have subwords $(12)(2)(12)$, by Lemma 7.6. Thus, in any case, subwords $(12)(2)(12)$ do not occur in br(h). Consider two subcases.

Case 1.2.1.1.1. Suppose that B does not contain R_p for $p \neq 1$. Then br(h) contains neither (3) nor (5) because $U \bullet h$ is a ring computation. This implies that br(h) cannot be longer than the words $(23)(2)(12)(1)(12)(2)(23)$ or $(51)(1)(12)(2)(23)$, $|\text{br}(h)| \leq 7$.

Case 1.2.1.1.2. Suppose that B contains R_p, $p \neq 1$. If B has no subwords $R_p\overrightarrow{R_p}$, then letters $(1), (12), (2), (23), (51)$ do not occur in br(h) (the locking argument again). Recall that by Case 1.2.1.1, (45) does not occur in br(h).

If B has a subword $R_p\overrightarrow{R_p}$, $p \neq 1$, then br(h) cannot contain a subwords $(51)(5)(51)$ or $(23)(3)(23)$ by Lemma 7.6. In view of the previous restrictions, the longest possible brief history cannot be longer than the word $(3)(23)(2)(12)(1)(12)(2)(23)(3)$, $|\text{br}(h)| \leq 9$.

Case 1.2.1.2. Suppose that there exists a subword $L_jP_jR_j$ in B, $j \neq 1$. Then br(h) has no subwords $(23)(2)(23)$, $(34)(3)(34)$, $(34)(4)(34)$ and $(45)(5)(45)$ by Lemma 7.6. The subwords $(12)(2)(12)$ and $(45)(4)(45)$ can be excluded as in Case 1.2.1.1. The subwords $(23)(3)(23)$ and $(51)(5)(51)$ are impossible by a similar reason (but one has to consider the alternative whether $R_j\overrightarrow{R_j}$ occurs in the base or not).

The restrictions we have obtained show that there is one of the letters $(12), (51)$ in br(h) if the length of the brief history is at least 8. Therefore B has a subword $\overleftarrow{L_j}L_jP_jR_j$ by the locking argument. Similarly, if the length of br(h) is at least 6, it has one of the letters $(12), (23), (51)$, and so B contain subword $\overleftarrow{L_j}L_jP_jR_j\overrightarrow{R_j}$. Continuing the extension of the subword of the base using the locking argument and Lemma 6.2, we see that the computation is tame if the length of br(h) is at least 8, contrary the assumption of the lemma.

Case 1.2.2. Let $yz \equiv P_jR_j$. Then by Lemma 6.2, one can assume that B has a subword $P_j^{-1}P_jR_j$ (otherwise we have Case 1.2.1.2). Hence br(h) has neither of the letters $(23), (3), (34)$ (the locking argument). This brief history cannot contain subwords $(51)(5)(51)$ by Lemma 7.6.

If there are no letters L_p in B for $p \neq 1$, then there are no letters (2), (4) in br(h) because rules from $\bar{S}(2) \cup \bar{S}(4)$ are not active with respect to zz'-sectors if $z, z' \neq L_p$. Otherwise there are no subwords $(12)(2)(12)$ or $(45)(4)(45)$ in br(h) by Lemma 7.6. Therefore such subwords are absent in any case. Hence br(h) cannot be longer than words $(2)(12)(1)(51)(5)(45)(4)$, $(2)(12)(1)(12)(2)$, $(4)(45)(5)(51)(1)(51)(5)(45)(4)$, $|\text{br}(h)| \leq 9$.

Case 2. Suppose now that B does not contain P_j for $j \neq 1$. Since the computation is a ring computation, br(h) does not contain (1).

If there are nether L_j- nor R_j-letters in B for $j \neq 1$, then br(h) contains no letters from $\{(1),...,(5)\}$. Therefore br(h) must be a subword of a product of several factors of the form $(12)(23)(34)(45)(51)$ or $(51)(45)(34)(23)(12)$ because the history of the computation is reduced. However br(h) cannot contain two occurrences of (12) or (51) otherwise the computation is not a ring computation. Indeed the transition rules not from $\mathcal{S}(34)$ change only $(\bar{\mathcal{E}}, \Omega)$-coordinates of the words, rules from $\mathcal{S}(34)$ also change only $(\bar{\mathcal{E}}, \Omega)$-coordinates in admissible words without L_j in their bases. Hence if br(h) contains two occurrences of (12) or (51), the corresponding ring is not reduced. Thus, the length of br(h) is at most 13 in this case. Hence, one may assume further that there is either L_j- or R_j-letter in B for some $j \neq 1$.

Case 2.1. Suppose that B contains L_j, $j \neq 1$. Then the locking argument shows that br(h) contains no letters from $\{(23),(3),(34)\}$. Then the standard application of Lemma 7.6 shows that br(h) cannot contain either $(12)(2)(12)$ or $(45)(4)(45)$.

If B has no letter R_j with $j \neq 1$ then br(h) has no letter (5) (otherwise the computation is a ring computation since the corresponding diagram is compressible). Hence by Lemma 7.6, br(h) has no subwords $(51)(5)(51)$ whether B contains R_j or not. Hence br(h) cannot be longer than the words

$$(2)(12)(1)(51)(5)(45)(4), (2)(12)(1)(12)(2), (4)(45)(5)(51)(1)(51)(5)(45)(4),$$

$|\text{br}(h)| \leq 9$.

Case 2.2. Suppose that B contains R_j, $j \neq 1$, but L_p does not occur in B if $p \neq 1..$ Then br(h) cannot contain letters $(34),(4),(45)$ since $P_j R_j$ is not a subword of B. Besides br(h) has no letters $(2),(4)$ since otherwise the computation is not a ring computation. The subwords $(23)(3)(23)$, $(51)(5)(51)$ are impossible by Lemma 7.6. Hence br(h) cannot be longer than the words $(3)(23)(12)(1)(51)(5), (3)(23)(12)(1)(12)(23)(3), (5)(51)(1)(51)(5), |\text{br}(h)| \leq 7$.

The lemma is proved. $\square$

Lemma 7.26. *There exists an algorithm to check, given two admissible words U, V for $\bar{\mathcal{S}}$, if there exist a ring computation $U \bullet h = V$ of $\bar{\mathcal{S}}$. The length of $|h|$ is recursively bounded in terms of $|U| + |V|$.*

Proof. We can assume that the bases of U and V are the same, the first and the last letters of that base coincide.

First we check, using Lemma 7.24, if both words are acceptable by $\bar{\mathcal{S}}$.

Case 1. Suppose that the answer is "yes". Let $U \bullet h_1 = \Sigma(w_1)^s K_1(\emptyset, 1)$ and $V \bullet h_2 = \Sigma(w_2)^s K_1(\emptyset, 1)$ be accepting computations.

Let δ be the homomorphism $\mathcal{H} \to \bar{\mathcal{G}}$ from Lemma 3.1. By Lemma 3.7 and Lemma 6.9, if $\delta(w_1)$ and $\delta(w_2)$ are equal in $\bar{\mathcal{G}}$ then there exists a computation $\Sigma(w_1)^s K_1(\emptyset, 1) \bullet h_3 = \Sigma(w_2)^s K_1(\emptyset, 1)$. Then $U \bullet h_1 h_3 h_2^{-1} = V$ (or another computation if the corresponding annular diagram is not reduced) is a ring computation (by Lemma 6.16).

Suppose that $\delta(w_1) \neq \delta(w_2)$ in $\bar{\mathcal{G}}$. Then by Lemma 3.2 there exist no computations connecting $\Sigma(w_1)^s K_1(\emptyset, 1)$ and $\Sigma(w_2)^s K_1(\emptyset, 1)$. Hence there is no computation connecting U and V. Since the word problem in $\bar{\mathcal{G}}$ is decidable, we deduce that, in Case 1, we can check if U and V are connected by a computation of $\bar{\mathcal{S}}$.

Case 2. Suppose that U or V are not accepted by $\bar{\mathcal{S}}$. Then we can assume that both are not accepted (otherwise U and V obvious cannot be connected by a computation). Then by Lemmas 7.25 and 7.24, one has to check only computations whose brief histories have lengths at most 13 (if U is wild) or 19 if U is tame. This can be done by Lemma 7.23.

In both cases we obtain a computation (if it exists) of recursively bounded length. Thus the number of transition rules is bounded for every ring computation which connect U and V. Then Lemma 7.23 helps us to effectively select a desired ring computation. □

7.5 All ring computations

Lemma 7.27. *Let $U \bullet h = V$ be a ring computation of $\mathcal{S} \cup \bar{\mathcal{S}}$. Then*

(1) $h = h_1 h_2 h_3$ where h_1, h_3 are words in $\mathcal{S}$, h_2 is a word in $\bar{\mathcal{S}}$.

(2) If h_2 is not empty, then $|br(h_1)|, |br(h_3)|$ are recursively bounded in terms of $|U|, |V|$.

Proof. (1) Suppose $h = h' h_0 h''$ where non-empty word h_0 is over $\mathcal{S}$, h' ends and h'' starts with a rule from $\bar{\mathcal{S}}$. Let $U' = U \bullet h'$, and consider the computation $U' \bullet h_0$. Denote by Δ the ring corresponding to the computation $U \bullet h$. Notice that a rule of $\mathcal{S}$ as well as a rule from $\bar{\mathcal{S}}$ can be applied to U' and to $U' \circ h_0$. Hence there are no a-letters in U' except for $a(P_j)$-letters. Therefore the subring Δ_0 of Δ, which correspond to $U' \bullet h_0$ has no $\mathcal{X}$-letters by Lemma 6.1.

Notice that by definition of $\bar{\mathcal{S}}$ if in every rule τ of $\mathcal{S}$, we remove a-letters in subrules $z \to uz'v$ where $z \in \{K_1, L_1, P_1, R_1\}$ and put ¯ over every letter of the rule, we get a rule $\bar{\tau}$ from $\bar{\mathcal{S}}$. Moreover, if for any admissible word W in the domain of τ, we remove a-letters from zz'-sectors where $z, z' \in \{K_1, L_1, P_1, R_1, K_2\}$, and add ¯ to every letter, we get an admissible word $\bar{W}$ for $\bar{\mathcal{S}}$, and
$$\bar{W} \circ \bar{\tau} = \overline{W \circ \tau}.$$

Since the words U' and $U' \circ h_0$ are in domains of rules from both $\bar{\mathcal{S}}$ and $\mathcal{S}$, $\overleftarrow{L_j}z$, $L_j z$-, $R_j z$-sectors of these words do not contain a-letters, for any j, the $\overleftarrow{L_1}$-, $L_1 z$-, $P_1 z$-, $R_1 z$-sectors do not contain a-letters, the a-letters in $P_j z$-sectors are from $\{a_1(P_1), ..., a_m(P_1)\} = \{\bar{a}_1(P_1), ..., \bar{a}_m(P_1)\}$, and the $(\bar{\mathcal{E}}, \Omega)$-coordinates of these words are $(\emptyset, 1)$. Hence $\bar{U'} = U$ and $\overline{U' \circ h_0} = U' \circ h_0$. Therefore we can replace h_0 by a word $\bar{h}_0$ over $\bar{\mathcal{S}}$ so that $U' \circ \bar{h}_0 = U' \circ h_0$. By

Lemmas 6.16 and 7.2 and the absence of $\mathfrak{X}$-edges in the boundary of Δ_0, the computation $U \bullet h'\bar{h}_0 h'' = V$ or another computation if the corresponding ring diagram is not reduced, is a ring computation, but some Θ-letters of h are replaced by $\bar{\Theta}$-letters in $h'\bar{h}_0 h''$. This contradicts the definition of reduced and compressible diagrams, because (Θ, k)-cells are higher than $(\bar{\Theta}, k)$-cells.

(2) Assume h_2 is non-empty and consider, for instance, $\mathrm{br}(h_1)$. If U is wild and there are no letters (34) in $\mathrm{br}(h_1)$, then the forms (i) - (iv) from the proof of Lemma 7.19 show that $|\mathrm{br}(h_1)| \leq 9$. If U is wild, and $\mathrm{br}(h_1)$ contains (34) and one of the letters $(1), (12), (2), (23), (45), (5), (51)$, then, by Lemma 7.16, we can assume that $U^{\pm 1}$ contains an admissible subword with base $\overleftarrow{L_j} L_j P_j R_j \overrightarrow{R_j}$ for some $j \neq 1$. If U is wild, and every letter of $\mathrm{br}(h_1)$ belongs to $\{(3), (34), (4)\}$, then $|\mathrm{br}(h_1)| \leq 5$ as was shown in the proof of Lemma 7.18. Thus, we may assume further that the base of $U^{\pm 1}$ contains an admissible subword with base $\overleftarrow{L_j} L_j P_j R_j \overrightarrow{R_j}$ for some $j \neq 1$.

By Lemma 7.7, we may assume that $\mathrm{br}(h_1)$ contains each of the letters $(12), (23), (34), (45), (51)$. Using the locking argument as in Lemma 7.17, we conclude that U is tame. Then we may assume by Lemma 7.24(1), that U is accepted by $\mathcal{S}$: $U \circ g_1 = \Sigma^s(w) K_1(0, 1)$ for some ring computation $U \bullet g_1$ of $\mathcal{S}$.

Let U_1 be an admissible subword of the tame word $U^{\pm 1}$ with base $K_1 L_1 P_1 R_1 K_2^{-1}$. Then the standard application of the homomorphism δ (introduced before Lemma 3.1) to $U_1 \bullet g_1$ shows that $\delta(U_1) = \delta(w)$ in the group $\bar{\mathcal{G}}$. The same homomorphism applied to $U_1 \bullet h_1$ shows that $\delta(U_1) = 1$ in $\bar{\mathcal{G}}$ since the $\bar{\mathcal{S}}$-rules of h_2 lock $K_1 L_1$-, $L_1 P_1$-, $P_1 R_1$- and $R_1 K_2^{-1}$-sectors. Hence the positive word w represents the identity of $\bar{\mathcal{G}}$.

Now, by Lemmas 2.1 and 3.6, there exists a ring computation $(U \circ g_1) \bullet g_2 = \tilde{\Sigma}$ of $\mathcal{S}$. Set $g = g_1 g_2$, then the length of the ring computation $U \bullet g = \tilde{\Sigma}$ (or of some other ring computation if the annular diagram, corresponding to $U \bullet g$ is not reduced) is recursively bounded as function of $|U|$. (Since we already know that such a computation exists, and the number of its $\mathcal{S}(34)$-rules is bounded, we get an upper bound for its length as in the proof of Lemma 7.21(2).)

The computation $U \bullet gg^{-1} h_1$ corresponds to a diagram consisting of three reduced rings. It contains the subcomputation $\tilde{\Sigma} \bullet (g^{-1} h_1) = U \circ h_1$. Since $\tilde{\Sigma} \equiv \tilde{\Sigma}$ and $U \circ h_1$ is in the domain of h_2, we have, as in part (1) of the proof, a ring computation of $\bar{\mathcal{S}}$: $\tilde{\Sigma} \bullet f = U \circ h_1$. Hence the ring computation $U \bullet h'$ can be replaced by the computation $U \bullet g(fh_2) h_3$ with recursively bounded length of $\mathrm{br}(g)$ where computation $(U \circ g) \bullet fh_2$ is a computation of $\bar{\mathcal{S}}$). Therefore the number of rules from $\mathcal{S}(34)$ is also recursively bounded in h_1 since otherwise Δ were compressible. Then, by Lemma 7.7, we have a recursive upper bound for the length of $\mathrm{br}(h_1)$. (Again $(U \circ g) \bullet fh_2$ can be replaced by a ring computation if the corresponding annular diagram is not reduced.) $\square$

According to Lemma 7.27, we will consider ring computations of the form $U \bullet h_1 h_2 h_3 = V$ where h_1, h_3 are words over $\mathcal{S}$, h_2 is a word over $\bar{\mathcal{S}}$, and if h_2 is not empty, then $|\mathrm{br}(h_1)|, |\mathrm{br}(h_3)|$ are recursively bounded in terms of $|U|$ and $|V|$. We fix this notation for the next two lemmas.

Lemma 7.28. *Assume the word h_2 is non-empty. Then the words h_1 and h_3 have lengths bounded by a recursive function of $|U|$ and $|V|$ respectively.*

Proof. Consider the computation $U \bullet h_1$ (the computation $V \bullet h_3^{-1}$ is similar).

Case 1. First suppose that the base of the computation contains the j-standard base $\overleftarrow{L}_j L_j P_j R_j \overrightarrow{R}_j$, $j \neq 1$. Then we can consider the restriction of the computation to a subword of U with the j-standard base. We know that the length of $\mathrm{br}(h_1)$ is recursively bounded. Therefore the number of occurrences of (34) in h is also bounded. Notice also that since $U \circ h_1$ is in the domain of a rule of $\bar{\mathcal{S}}$, the $\overleftarrow{L}_j L_j$-, $L_j P_j$, $R_j \overrightarrow{R}_j$-sectors of $U \circ h_1$ do not contain a-letters. Therefore, by Lemma 7.10(b), $|\overleftarrow{L}_j(U \circ h_1[t])| + |L_j(U \circ h_1[t])| + |R_j(U \circ h_1[t])|$, $t = 1, \ldots, |h_1|$ is bounded by a recursive function in $|U|$. Then Lemma 7.5 gives a recursive upper bound for lengths of arbitrary maximal subwords of h_1 over $\mathcal{S}(i)$, $i = 1, \ldots, 5$. Since the number of such subwords is recursively bounded, we get a recursive bound for $|h_1|$ in terms of $|U|$ and the claim is proved.

Case 2. Assume now that the base of $U^{\pm 1}$ does not contain $W_j = \overleftarrow{L}_j L_j P_j R_j \overrightarrow{R}_j$ as a subword for any $j \neq 1$.

Notice, however, that since $U \circ h_1$ is in the domain of a rule from $\mathcal{S}$ and a rule from $\bar{\mathcal{S}}$, the word $U \circ h_1$ does not contain letters from $\mathcal{A}(K_j) \cup \bar{\mathcal{A}}(K_j) \cup \mathcal{A}(L_j) \cup \bar{\mathcal{A}}(L_j) \cup \mathcal{A}(R_j) \cup \bar{\mathcal{A}}(R_j)$ since such letters does not belong to $\mathcal{A} \cap \bar{\mathcal{A}}$. Therefore, by Lemma 2.2, every $\overleftarrow{L}_j$-, L_j-, P_j-, R_j- and $\overrightarrow{R}_j$-letter in the base of U must occur in subwords of the form $(\overleftarrow{L}_j L_j)^{\pm 1}$, $(R_j \overrightarrow{R}_j)^{\pm 1}$, $(L_j P_j)^{\pm 1}$. Hence, by Lemma 6.2 the base of the corresponding ring Δ must contain a letter L_j or a letter R_j for $j \neq 1$.

Let us consider only the first case, because one can apply the same argument to the second one. In this case, the subword $(P_j R_j)^{\pm 1}$ cannot be a neighbor in the base to the occurrence of $L_j^{\pm 1}$, since otherwise we get a subword W_j in the base of U. Consequently, $\mathrm{br}(h_1)$ does not contain the letter (34) (because rules from $\mathcal{S}(34) \cup \bar{\mathcal{S}}(34)$ lock $L_j P_j$- and $P_j R_j$-sectors) and the base of $U^{\pm 1}$ contains a subword $W_j' = \overleftarrow{L}_j L_j P_j P_j^{-1} L_j^{-1} (\overleftarrow{L}_j)^{-1}$. Let U' be a subword of U with base W_j'.

Consider the computation $U' \bullet h_1$. The word $U' \circ h_1$ has the form

$$(\overleftarrow{L}_j L_j P_j v P_j^{-1} L_j^{-1} \overleftarrow{L}_j^{-1})(r, i)$$

$((r, i)$ are the $(\bar{\mathcal{E}}, \Omega)$-coordinates), where v is a word in $\{a_1(P_j), \ldots, a_m(P_j)\}$. (Recall that $a_i(P_j)$ is identified with a $\bar{a}_i(P_j)$ only if $i \leq m$.)

Let γ be the homomorphism defined in the proof of Lemma 7.13. Then $|\gamma(v)| \leq |U'|$. This means that we have a linear bound of $|U' \circ h_1|$ in terms of $|U'|$. Since this argument works for every subword of $U^{\pm 1}$ with base W'_j, $j \neq 1$, and $U \circ h_1$ contains no a-letters except letters from $\mathcal{A}(P_j) \cup \bar{\mathcal{A}}(P_j)$, $j \neq 1$, we obtain a linear bound for $|U \circ h_1|$ in terms of $|U|$. It remains to apply Lemma 7.20. □

Lemma 7.29. *There is an algorithm to check whether, given two admissible words U, V for $\mathcal{S} \cup \bar{\mathcal{S}}$, there exists a ring computation $U \bullet h = V$. The length of the history h is recursively bounded in terms $|U|, |V|$ for arbitrary such ring computation.*

Proof. By Lemma 7.22, we can assume that there are no computations of $\mathcal{S}$ connecting U and V. By Lemma 7.26, we can assume that there is no ring computation of $\bar{\mathcal{S}}$ connecting U and V. (Recall that all ring computations $U \bullet f = V$ of $\mathcal{S} \cup \bar{\mathcal{S}}$ must have equal length $|f|$ depending on U and V, by the definition of a reduced diagram.)

Hence, by Lemma 7.27, we have to analyze only the "mixed" situation when h_2 is not empty and h_1 or h_3 is not empty (see the notations introduced before Lemma 7.28).

The sum $|h_1|+|h_3|$ is recursively bounded by Lemma 7.28. Therefore the lengths $|U \circ h_1|$ and $|U \circ h_1 h_2|$ are recursively bounded modulo $\mathcal{G}$-relations (i.e., if we substitute the subwords in $\mathcal{A}(P_1)$ by equal modulo $\mathcal{G}$ shorter subwords). Therefore by Lemma 7.26, we can recursively bound $|h_2|$. Hence we can recursively bound the sum $|h| = |h_1| + |h_2| + |h_3|$. □

Lemma 7.30. *Let Δ be any annular diagram over Δ whose contours p, q do not contain θ-edges. Then there exists another annular diagram Δ' with the same boundary labels and recursively bounded number of cells (in terms of $|p| + |q|$). Moreover, Δ' and Δ have equal histories if Δ is a ring.*

Proof. By Lemma 6.4, we can assume that Δ is a ring or a quasiring. Lemma 6.14 treats the quasiring case, so we can assume that Δ is a ring. By Lemma 7.29, we can bound the size of the corresponding ring computation. Finally Lemma 7.1 bounds the number of cells in Δ. □

8 Spirals

We need the following properties of words in a free group F.

Lemma 8.1. *For arbitrary elements u, v, w of F and any integer $d \geq 0$, the length of an arbitrary product $u^t w v^t$ in F is not greater than $2(|u|+|v|+|w|) + |u^d w v^d|$ provided $0 \leq t \leq d$. Furthermore, either equality $u^t w v^t = w$ is true in F for every t or $|u^t w v^t| \geq t - 2(|u| + |v| + |w|)$.*

Proof. Since, for any word U, a power U^s is freely equal to $Y U_0^s Y^{-1}$ where U_0 is the cyclically reduced form of U and $|Y| \leq |U|$, it suffices to assume that the words u and v are cyclically reduced. Moreover one of them can be assumed to be non-empty, and words u and v can be assumed to be not equal to proper powers of some non-trivial elements of F.

Of course, there may be cancellations in the words $u^t w v^t$.

Suppose first, that the cancellations are not substantial in the following sense: the length of the suffix (resp. prefix) of $u^t w$ (resp $w v^t$) that is cancelled in the product $u^t w v^t$ does not exceed $2(|w| + |u| + |v|)$ for every t. Then the sequence of lengths $|u^t w v^t|$ increases linearly with the slope $\geq |u| + |v| \geq 1$ for $t \geq 2(|w| + |u| + |v|)$. Hence $|u^t w v^t| \leq 2(|u| + |v| + |w|) + |u^d w v^d|$, as desired.

Now suppose that there exists a substantial cancellation in the above mentioned sense. Then both words u^t and v^{-t} contain a common subword T with $|T| \geq |u| + |v|$. This implies (see, e.g. [FW]) that u and v^{-1} coincide up to a cyclic shift: $v^{-1} \equiv u_2 u_1$ and $u \equiv u_1 u_2$ for some words u_1, u_2. Furthermore the word $w u_1^{-1}$ must be freely equal to to a power u^l for some l. Therefore $u^t w v^t = u^t u^l u^{1-t} u_2^{-1} = u^{l+1} u_2^{-1} = w u_1^{-1} u_1 u_2 u_2^{-1} = w$ in F for arbitrary t. The lemma is proved. □

By Britton's lemma, every x-letter has infinite order in the group $\mathcal{H}_2$. It is also convenient to introduce auxiliary letters $x^{1/4}, x^{1/16}, \ldots$ for every letter $x \in \mathcal{X}$. In other words, we embed our group $\mathcal{H}_2$ into the multiple amalgamated product of $\mathcal{H}_2$ and infinite cyclic groups $C_{x,i} = \langle x^{\frac{1}{4^i}}, x \in \mathcal{X} \rangle, i = 0, 1, 2, \ldots$ where $C_{x,0} = \langle x \rangle$, $x \in \mathcal{X}$, and for every $i \geq 1$, $C_{x,i}$ is identified with the subgroup of $C_{x,i+1}$ generated by $(x^{\frac{1}{4^{i+1}}})^4$. Letters $x^{\frac{1}{4^n}}$ will be called *fractional letters*, words in these letters will be called *fractional words*.

Lemma 8.2. *Let Δ be a quasitrapezium with the standard decomposition of the boundary $p_1 p_2 p_3 p_4$. Assume that the height of Δ is 2, Δ has no $\mathcal{G}$-cells and w_1, w_3 are maximal subwords in the alphabet $\mathcal{X}$ of $\phi(p_1), \phi(p_3^{-1})$, respectively. (Recall that $p_1 = \mathbf{top}(\mathcal{B})$, $p_3^{-1} = \mathbf{bot}(\mathcal{B}')$ as in the definition of trapezium or quasitrapezium.) Then w_3 is freely equal to $u w_1' v$ where w_1' is obtained from w_1 by an injective substitution of the form $x \mapsto (x')^{4^c}$ $(x, x' \in \mathcal{X})$, the integer c (same for all $x \in \mathcal{X}$) and the fractional words u, v depend recursively only on the boundary labels of the maximal θ-bands $\mathcal{T}_1$, $\mathcal{T}_2$ of Δ.*

Proof. Consider the subdiagram Γ of Δ bounded by $\mathbf{top}(\mathcal{T}_1)$ and $\mathbf{bot}(\mathcal{T}_2)$. Let $\bar{p}_1 \bar{p}_2 \bar{p}_3 \bar{p}_4$ be its boundary, where $\phi(\bar{p}_1) \equiv w_1$ and $\phi(\bar{p}_3)^{-1} \equiv w_3$. Consider the maximal system of maximal z-bands $\mathcal{B}_1, \ldots, \mathcal{B}_n$ which start on $\bar{p}_4$ and end on $\bar{p}_2$, where z is one of k- or a-letters. In particular, $\mathcal{B}_1$ and $\mathcal{B}_n$ are subbands of $\mathcal{B}$ and $\mathcal{B}'$ respectively. Let $p_1^i p_2^i p_3^i p_4^i$ be the standard decomposition of the boundary $\partial(\mathcal{B}_i)$, where p_2^i, p_4^i are edges from the paths $\bar{p}_2$ and $\bar{p}_4$, respectively, and the labels $w_1^{(i)}, w_3^{(i)}$ of $p_1^i, (p_3^i)^{-1}$ are some words in $\mathcal{X}$.

In particular, $w_1^{(1)} \equiv w_1$ and $w_3^{(n)} \equiv w_3$. The words $w_1^{(i)}, w_3^{(i)}$ are empty if $\mathcal{B}_i$ is a P_j- or R_j-band, since there are no (P_j, x)- or (R_j, x)-relations of the form (2.9).

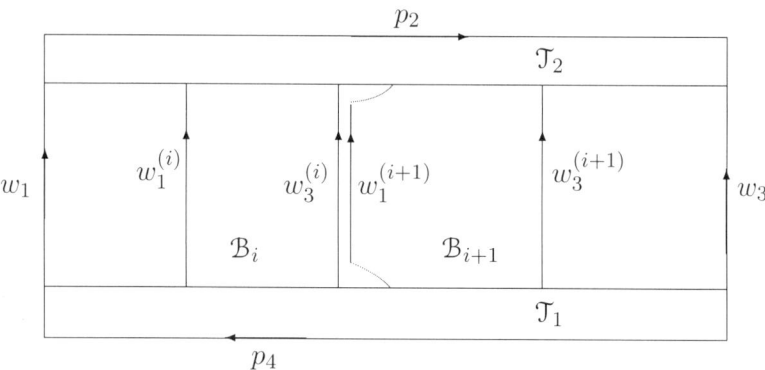

Fig. 20.

Notice, that by x-relations (2.8) and (2.9), the word $w_3^{(i)}$ is obtained from the word $w_1^{(i)}$ by an injective substitution of the form $x \mapsto (x')^l$ on the set of x-letters of the word $w_1^{(i)}$ where $l \in \{1, 4, 1/4\}$, l does not depend on an x-letter x.

There are no cells of Γ in the maximal subdiagram of Δ whose boundary consists of x-edges, and which is situated between $\mathcal{B}_i$ and $\mathcal{B}_{i+1}$. Indeed, by Lemma 3.11, otherwise any non-$\mathcal{G}$-cell would be included in an annulus, and there are at most two edges of $\bar{p}_2$ (x-edges) between p_2^i and p_2^{i+1} and at most 2 edges of $\bar{p}_4$ between p_4^i and p_4^{i+1} (it follows from Lemma 6.1). Hence the word $w_1^{(i+1)}$ is freely equal to $u_i w_3^{(i)} v_i$ for $|u_i|, |v_i| \leq 2$, and $w_1^{(i+1)}$ is obtained from from $w_1^{(i)}$ by the above mentioned substitution $x \mapsto (x')^l$ and multiplication by the words u_i, v_i which depend only on the bands $\mathcal{T}_1, \mathcal{T}_2$. The obvious induction on $i \leq n$, shows that w_3 is freely equal to $uw_1' v$ where w' is obtained from w_1 by a substitution of the form $x \mapsto (x')^{4^c}$ with $|c| \leq n$ (c depends on the bands $\mathcal{T}_1, \mathcal{T}_2$), and u, v are fractional words which can be effectively calculated given $\mathcal{T}_1, \mathcal{T}_2$. □

Let Δ be a trapezium with at least 2-letter base. Suppose that the side bands $\mathcal{B}$ and $\mathcal{B}'$ of Δ are partitioned into subbands $\mathcal{B}_1, \mathcal{B}_2$ and $\mathcal{B}_1', \mathcal{B}_2'$ respectively, such that the band $\mathcal{B}_1$ is the copy of $\mathcal{B}_2'$ (or $\mathcal{B}_2$ is the copy of $\mathcal{B}_1'$), and each of the subbands $\mathcal{B}_1, \mathcal{B}_2, \mathcal{B}_1', \mathcal{B}_2'$ has a positive number of (θ, k)-cells.

We can identify $\mathcal{B}_1$ with $\mathcal{B}_2'$ (or $\mathcal{B}_2$ with $\mathcal{B}_1'$). If the resulting annular diagram Γ is reduced and has minimal boundaries, it is called a *spiral* (see Figure 21).

THE CONJUGACY PROBLEM AND HIGMAN EMBEDDINGS 89

Thus, a spiral Γ contains a k-band $\mathcal{B}''$, which connects its inner contour q with the outer contour p; this band $\mathcal{B}''$ is obtained from $\mathcal{B}$ and $\mathcal{B}'$ by the partial identification. The k-band $\mathcal{B}''$ will be called the *main k-band* of the spiral. Notice that the length of the identified portion of the bands $\mathcal{B}$ and $\mathcal{B}'$ does not exceed the lengths of the contours of the spiral. Our goal will be to estimate the number of cells in a spiral with the given boundary labels in terms of the lengths of the contours.

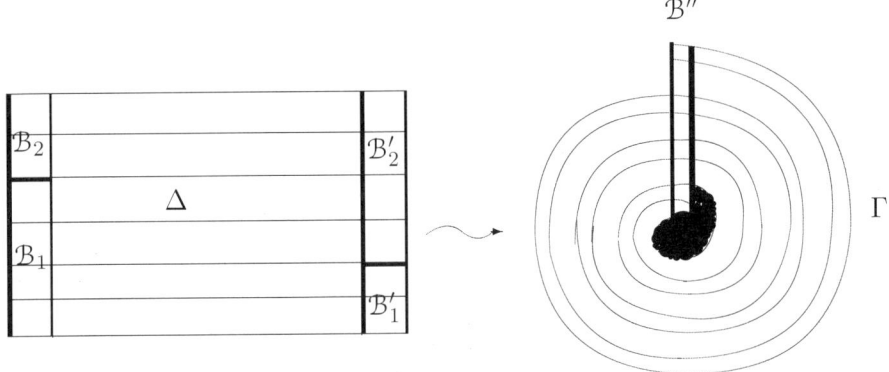

Figure 21.

It is easy to see that every maximal θ-band of Γ is a union of θ-bands of Δ, it starts on the inner contour q of Γ and ends on the outer contour p. It may intersect the k-band of Γ many times, and looks like a spiral. (However it can intersect every k-band of Γ only in one direction: this follows from Lemma 3.11.) Hence the number of maximal θ-bands of the spiral is at most $\min(|q|, |p|)$.

Let h_1, h_2 be the histories of the bands $\mathcal{B}_1$, $\mathcal{B}_2$. Then $\mathcal{B}$ and $\mathcal{B}'$ have the history $h_1 h_2$. On the other hand h_1 is a suffix of the history of $\mathcal{B}'$. Hence $h_1 h_2 = h_3 h_1$ for some h_3. If $|h_1| > |h_2|$, then this immediately implies that $h_1 h_2$ is a periodic word with period h_2 (i.e. $h_1 h_2$ is a subword of h_2^l for some $l > 0$). Notice that $|h_2|$ does not exceed the length of the inner contour of the spiral. So $|h_1 h_2| < 2|h_2| \leq |q| + |p|$ if $|h_1| \leq |h_2|$, because every θ-band in Γ connects q with p. Hence in any case, the history of Δ is a periodic word whose period does not exceed $|q| + |p|$.

Quasispirals are defined like spirals when one takes a quasitrapezium with empty base as the original diagram instead of a trapezium.

Lemma 8.3. *Let Δ be a reduced annular diagram over $\mathcal{H}_1$ with minimal boundaries, whose labels contain at least one θ-letter and a k- or a-letter, $a \notin \mathcal{A}(P_1)$. Let the corresponding maximal k- or a-band $\mathcal{C}$ have recursively bounded (in terms of $|p|+|q|$) number of θ-cells. Then either (1) there exists a reduced annular diagram Δ' over $\mathcal{H}_1$ with the same boundary labels as Δ and with recursively bounded (in terms of $|p| + |q|$) number and perimeters* of cells, or (2) there is a reduced annular diagram Δ' without $\mathcal{G}$-cells and with minimal boundaries p' and q' which are at most twice longer than p and q respectively, and $\phi(p)$ and $\phi(p')$ (resp. $\phi(q), \phi(q')$) are conjugate and the number and lengths of relations needed to deduce these conjugacies are recursively bounded.

Proof. By Lemma 5.6, we can assume that Δ has minimal boundaries. If there exists an a-band starting on a $\mathcal{G}$-cell in Δ and ending on a k-band of Δ then this must be a P_1- or an R_1-band, and so we would be able to turn Δ into a simply connected van Kampen diagram by cutting along a side of this k-band. The length of this cut is recursively bounded as this follows from the lemma assumption and the lack of x-cells in P_1- and R_1-bands. After that we could apply Lemma 5.1. Thus we can assume that every a-band starting on a $\mathcal{G}$-cell of Δ ends on p or q. Therefore by Lemma 3.12, parts (1), (2), we can remove all the $\mathcal{G}$-cells from Δ increasing the length of the boundary at most by a factor of 2. The total perimeter of all $\mathcal{G}$-cells that we remove is bounded by $|p| + |q|$. □

Lemma 8.4. *Let Δ be a reduced annular diagram over $\mathcal{H}_1$ with minimal boundaries p and q having a θ-band $\mathcal{T}$ and at least one k- or a-edge for an a-letter $a \in (\mathcal{A} \cup \bar{\mathcal{A}}) \backslash \mathcal{A}(P_1)$. Let $\mathcal{C}$ be corresponding k or a-band connecting the contours q and p. Then there exists a diagram Δ' with the same boundary label as Δ, and such that $\Delta' = \Delta_1 \cup \Delta_2 \cup \Delta_3$ where the number and perimeters of cells in Δ_1, Δ_3 are recursively bounded and Δ_2 is either empty or a spiral, or a quasispiral.*

Proof. By Lemmas 5.4, 5.5, we can obtain a minimal diagram Δ' satisfying (R1). Then all maximal θ-bands in Δ connect p and q, so their number does not exceed $|p| + |q|$. Also all maximal k-bands in Δ' connect p and q. If there are no k-bands, then every a-band, for $a \notin \mathcal{A}(P_1)$, connects p and q.

All maximal a-bands starting on p or q end either on q and p respectively, or on a k-band or on a $\mathcal{G}$-cell (recall that different a-bands cannot cross and different k-bands cannot cross).

Suppose first that every θ-band in Δ intersects every k-band and every a-band, $a \notin \mathcal{A}(P_1)$, in Δ at most twice. Since the number of maximal θ-bands in Δ is bounded by $|p| + |q|$, the number of θ-cells in $\mathcal{C}$ is at most $2(|p| + |q|)$. By Lemma 8.3 we can assume that Δ does not contain $\mathcal{G}$-cells. Since the number of maximal k-bands in Δ is bounded by $|p| + |q|$, the number of θ-cells in k-bands of Δ is bounded by $2(|p| + |q|)^2$. Therefore the number of maximal a-bands in Δ is bounded by $2(|p| + |q|)^2 + (|p| + |q|) < 3(|p| + |q|)^2$. Since each of the a-bands intersects each of the θ-bands at most twice, the number of θ-cells in each of them is bounded by $2(|p| + |q|)$. Therefore the total number of θ-cells in Δ is bounded by $6(|p| + |q|)^3 + 2(|p| + |q|)^2$. Therefore the θ-band $\mathcal{T}$ has a recursively bounded length. Thus we can cut Δ along a side of $\mathcal{T}$, obtain a simply connected van Kampen diagram with

THE CONJUGACY PROBLEM AND HIGMAN EMBEDDINGS 91

perimeter recursively bounded in terms of $|p|+|q|$, and use Lemma 5.1 to bound the number of cells in Δ.

Notice that if a θ-band intersects an a- or k-band $\mathcal{C}'$ three times then by Lemma 3.11 it must go around the hole of the diagram twice and so by Jordan's lemma it intersects the a- or k-band $\mathcal{C}$ connecting p and q at least twice.

Thus we can assume that the θ-band $\mathcal{T}$ in Δ intersects $\mathcal{C}$ twice. Moreover we can conclude that there is a subband $\mathcal{C}_0$ of $\mathcal{C}$ with linearly (in terms of $|p|+|q|$) bounded number of θ-cells in $\mathcal{C}\setminus\mathcal{C}_0$, such that for every θ-cell π of $\mathcal{C}_0$ there is a θ-band $\mathcal{T}_\pi$ which starts with π and whose last cell π' (and no other cell) also belongs to $\mathcal{C}$, $\pi' \neq \pi$. These bands $\mathcal{T}_\pi$, for $\pi \in \mathcal{C}_0$, (and the auxiliary cells between them) form a subdiagram Δ_2 which is a spiral or a quasispiral with main band $\mathcal{C}_0$ by the construction and the assumptions on p and q. If there exists a $\mathcal{G}$-cell in Δ outside Δ_2 and an a-band connecting it with the boundary of Δ_2, we can use Lemma 3.12, move the $\mathcal{G}$-cell to the boundary of Δ_2, and then include that cell in Δ_2. Thus we can assume that there are no $\mathcal{G}$-cells in Δ outside Δ_2 which are connected with the boundary of Δ_2 by an a-band.

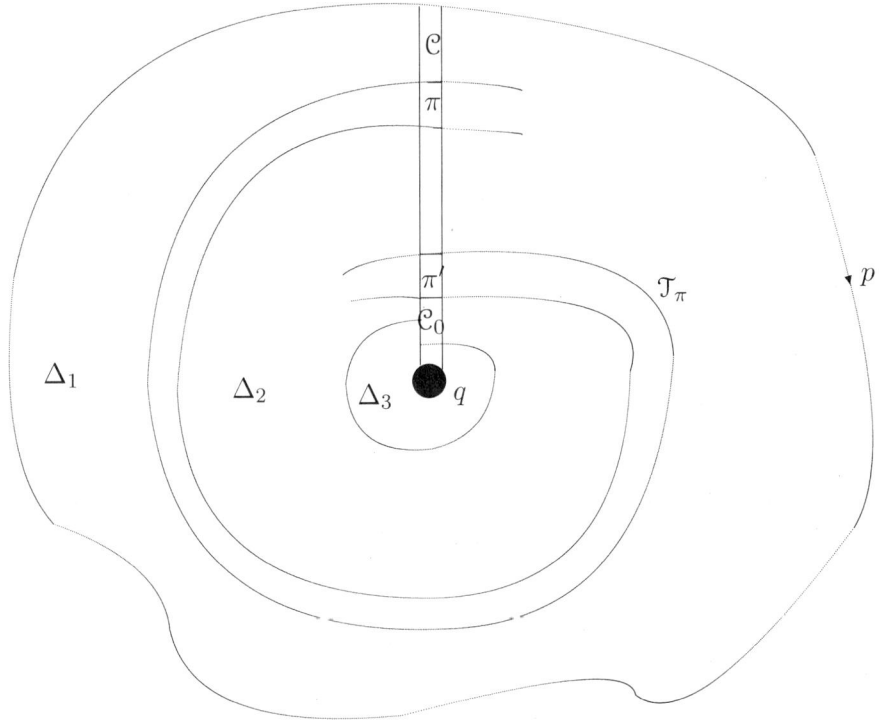

Fig. 22.

The diagram Δ is a union of Δ_2 and two annular diagrams Δ_1, Δ_3 which are situated between Δ_2 and p and between Δ_2 and q respectively.

Our purpose is to obtain a recursive upper bound for the number of cells in Δ_1 and Δ_3 as a function in $|p|, |q|$. We will consider only the diagram Δ_1 (Δ_3 is similar).

By the choice of $\mathcal{C}_0$, each of the maximal θ-bands in Δ has linearly bounded (in terms of $|p|+|q|$) number of intersections outside Δ_2 with every maximal a- or k-band of Δ. Notice also that an a-band of Δ_1 which has at least one θ-cell, cannot start and end on the boundary of Δ_2 by Lemma 3.11. Hence every maximal a-band of Δ_1 must start and end on p or on a (k,θ)-cell, and therefore the number of θ-cells in Δ_1 is recursively bounded in terms of $|p| + |q|$. By our assumption, every a-band starting on a $\mathcal{G}$-cell of Δ_1 ends on p or on a θ-cell of Δ_1. Therefore the total perimeter of all $\mathcal{G}$-cells in Δ_1 is recursively bounded.

Since θ-cells of Δ_1 are arranged in θ-bands starting on p, we may delete all (a recursively bounded number) of them from Δ_1 (the result will be again an annular diagram) and assume that every cell of Δ_1 is an (a, x)-cell or a (k, x)-cell corresponding to the relations (2.8), (2.9), or a $\mathcal{G}$-cell. Then the $\mathcal{G}$-cells can also be deleted from Δ_1 by Lemma 3.12 (and such transformations do not change the boundary of Δ_2).

Since there are no a-, k-, (a, θ)-, (k, θ)-annuli in Δ by Lemma 3.11, all cells of Δ_1 belong to maximal a- and k-bands $\mathcal{C}_1, \mathcal{C}_2, \ldots$ starting on p (and ending on the other boundary component of Δ_1). Hence it suffices to bound the length of each of these a- and k-bands in Δ_1.

Notice that the common boundary component, say u, of Δ_1 and Δ_2 is the boundary component of a spiral (or quasispiral). By the above construction of Δ_2, we have $u = u_1 u_2$ where u_1 belongs to the boundary of $\mathcal{C}_0$, the last edge of u_1 is a θ-edge, and the label of u_2 is the label of a θ-band (modulo $\mathcal{G}$).

Suppose that there is a band $\mathcal{C}_i$ among $\mathcal{C}_1, \mathcal{C}_2, \ldots$, which ends on u_2. We may assume that i is chosen so that the end of $\mathcal{C}_i$ is the nearest to the beginning of u_2 among all $\mathcal{C}_j$ ending on u_2. Then $\mathbf{bot}(\mathcal{C}_i)$ cannot have common edges with any $\mathcal{C}_j$ since the last edge of u_1 is a θ-edge, which also belong to p. (Recall that now we have no θ-cells in Δ_1.) The bottom of $\mathcal{C}_i$ has at most 2 common edges with u_2 because u_2 can have at most 2 consecutive x-letter in its boundary label by Lemma 6.1. Hence the length of $\mathcal{C}_i$ is recursively bounded by $|p|$.

Then suppose that there is a band $\mathcal{C}_i$ ending on u_1 (and so $\mathcal{C}$ is a k-band). It is clear from the form of (k, θ)-relations that there is a subpath evf of $u_1^{\pm 1}$ such that e is the end of C_i, f is a θ-edge, and v contains no two consecutive x-letters in its label. We may assume that no C_j ends on v. Then, as in the first case, the length of C_i is recursively bounded by p.

Thus, in any case, there is a band, say $\mathcal{C}_1$, whose perimeter and whose number of cells are linearly bounded in terms of $|p|$. We delete $\mathcal{C}_1$ from Δ_1 and then choose a band in the remaining subdiagram of Δ_1, whose perimeter and whose number of cells are recursively bounded as well. Then we delete

$\mathcal{C}_2$, etc. Since the number of a-, k-bands $\mathcal{C}_1, \ldots$ is recursively bounded, we conclude that there is a recursive bound for the number of cells in Δ_1, and the lemma is proved. $\square$

Remark 8.1. The proof of Lemma 8.4 shows that for every maximal k-band $\mathcal{C}$ in a spiral Δ, one can obtain another spiral Δ' with contours p' and q', such that (a) the words $\phi(p)$ and $\phi(p')$ ($\phi(q)$ and $\phi(q')$) are conjugate modulo relations used in Δ; (b) the number and lengths of relations used to deduce these conjugacies are recursively bounded in terms of $|p|+|q|$; (c) the set of θ-cells of the main band of Δ' is a subset of the set of cells of the band $\mathcal{C}$. Thus if the base of a spiral has letters P_j or R_j we shall always assume that the main k-band of the spiral is a P_j- or a R_j-band.

Lemma 8.5. *Let h be a periodic word in $\mathcal{S} \cup \bar{\mathcal{S}}$ with period h_0 and let $W \bullet h = W'$ be a free computation of $\mathcal{S} \cup \bar{\mathcal{S}}$. Then the number of different words among $W \circ h[t]$, $t = 1, \ldots, |h|$ is recursively bounded in terms of $|h_0| + |W| + |W'|$.*

Proof. By lemma 2.6, for every zz'-sector of W there exist two words u and v depending on h_0 whose size is linearly bounded in terms of $|h_0|$, such that the inner part of the corresponding zz'-sector in each $W \circ h_0^t$ is equal in the free group to $u^t W_{zz'} v^t$ where $W_{zz'}$ is the inner part of the zz'-sector of W, $t = 1, \ldots, s$. By Lemma 8.1, we can assume that s is big enough so that $s - |u| - |v| > |W'|$, so $u^t W_{zz'} v^t = W_{zz'}$ for every $t = 1, \ldots, s$. Hence $W = W \circ h_0 = W \circ h_0^2 = \ldots$. Replacing W by $W \circ h'$ for any prefix h' of h_0, we get that $W \circ h' = W \circ hh' = W \circ h^2 h' = \ldots$. Therefore the number of different words in the computation associated with Δ is recursively bounded in terms of $|W| + |W'|$ and $|h_0|$. $\square$

Lemma 8.6. *Let a minimal diagram Γ be a spiral obtained from a trapezium Δ. Then the minimum of lengths of all the maximal k-bands of Γ is bounded by a recursive function of $|q|+|p|$, where q, p are inner and outer contours of Γ.*

Proof. Let W be the projection of the label of the bottom side of the trapezium Δ corresponding to the spiral Γ onto $\mathcal{A} \cup \bar{\mathcal{A}} \cup \mathcal{K} \cup \bar{\mathcal{K}}$, and let h be the history of Δ corresponding to Γ. Notice that $h = h_0^s h'$ for some s and some prefix h' of h_0, and $|W|, |h_0| \leq |q| + |p|$. We need to bound s in terms of $|q|+|p|$. Indeed, suppose we have bounded s. Suppose the spiral contains a $\mathcal{G}$-cells. If there exists a-band starting on a $\mathcal{G}$-cell there ends on a P_1- or R_1-band, and the spiral has a cut of a recursively bounded length along a side of that band since it has no $\mathcal{X}$-cells. Thus we can assume, by Lemma 3.12, that all a-bands starting on $\mathcal{G}$-cells end on the boundary of the spiral, in which case the total perimeter of these cells is recursively bounded, and we can move these cells out of the spiral using Lemma 3.12. Therefore we can assume that the spiral does not contain $\mathcal{G}$-cells. Since the number of

k-cells is recursively bounded in terms of $|h|$ and $|q|+|p|$, we have that the number of maximal a-bands is bounded too. Hence the length of θ-bands in Δ are bounded as well, because every such a band has recursively bounded number of intersections with every a-band. Therefore there is a cut of Δ with recursively bounded length, which makes possible to apply Lemma 5.1.

Let us assume that $s > |q| + |p|$.

Let us enumerate k-bands in Γ clockwise: $\mathcal{B}_1, ..., \mathcal{B}_c$. Let $\mathcal{T}$ be a maximal θ-band of Γ. We can decompose $\mathcal{T}$ into several subbands $\mathcal{T}(1), \mathcal{T}(2), \ldots, \mathcal{T}(r_c)$ such that every $\mathcal{T}(l)$ (with possible exception $\mathcal{T}(1)$) starts with a cell of the band $\mathcal{B}_1$ and contains at most one common cell with each of the k-bands $\mathcal{B}_1, \ldots \mathcal{B}_c$. The bands $\mathcal{T}(2), \mathcal{T}(3), \ldots$ will be called the *turns* of $\mathcal{T}$. For every $l = 2, ...$ there is a path $t(l)$ going along $\mathcal{B}_1$, which connect the initial vertex of the top $p(l)$ of the turn $\mathcal{T}(l)$ with that of the top of of $\mathcal{T}(l+1)$. Notice that $p(l+1)_- = p(l)_+$. Denote by $q(l)$ the loop $t(l)p(l)^{-1}$ associated with $\mathcal{T}(l)$. Notice also that for every $l = 1, 2..$ words $\phi(p(l))_{\mathcal{A} \cup \bar{\mathcal{A}} \cup \mathcal{K} \cup \bar{\mathcal{K}}}$ coincides for some i, with the word $W \circ h[i]$ with omitted the last k-letter z. The label of $t(l)$ consists of at most $|h_0|$ θ-letters and a number of x-letters.

Consider several cases.

Case 1. There is a P_j- or R_j-letter in the base of Δ.

Then we assume that $\mathcal{B}_1$ is a P_j- or a R_j-band by Remark 8.1. We have the following alternatives.

Case 1.1. Suppose that the history of Δ contains a rule from $\bar{\mathcal{S}} \cup \mathcal{S}(34) \cup \mathcal{S}(4) \cup \mathcal{S}(45)$.

Then we can assume that $\mathcal{T}$ corresponds to one of these rules τ. Let $W_{zz'}$ be a sector of W which is not a $P_1 R_1$-, $P_1 P_1^{-1}$- or $R_1^{-1} R_1$-sector. Then the computation $W_{zz'} \bullet h$ is a free computation and, by Lemma 8.5, the number of different words among $W_{zz'} \circ h[i]$, $i = 1, 2...$ is recursively bounded in terms of $|q| + |p|$. Notice that the rule τ locks $P_1 R_1$-sectors. Hence the $P_1 z$- and zR_1-sectors of $\phi(p(l))_{\mathcal{A} \cup \bar{\mathcal{A}} \cup \mathcal{K} \cup \bar{\mathcal{K}}} z$ are of length 2. Therefore by Lemma 6.1, the number of different words among $\phi(p(l))$, $l = 1, 2, ...$, is recursively bounded.

Since the sides of P_j-, and R_j-bands do not contain x-edges, the lengths of the paths $t(l)$, $l = 1, 2, ...$ do not exceed $|h_0| \leq |q| + |p|$.

Therefore the number of different labels of the loops $t(l)p(l)^{-1}$, $l = 1, 2, ...$ is recursively bounded in terms of $|q| + |p|$. Notice that for different l's these loops do not intersect each other. Since Γ is minimal, it cannot contain two non-intersecting loops with the same labels. Hence $|h|$ is recursively bounded in terms of $|q| + |p|$. Hence the length of $\mathcal{B}_1$ is recursively bounded, as desired.

Case 1.2. Suppose that no rules from $\bar{\mathcal{S}} \cup \mathcal{S}(34) \cup \mathcal{S}(4) \cup \mathcal{S}(45)$ occur in h.

The history h is reduced by Lemma 6.8. By Lemma 6.12, then $W \bullet h$ is a computation of $\mathcal{S}$ and W is an admissible word for $\mathcal{S}$. Then one can apply Lemma 7.3 to the computation $W \bullet h = W' \pmod{\mathcal{G}}$ corresponding to the trapezium Δ and obtain a free computation $W \bullet h = W''$ where $W'' = W'(\mod \mathcal{G})$. Notice any two P_j-bands (resp. R_j-bands) which have the same histories consist of the same cells since there are no (P_j, x)- and (R_j, x)-relations of the form (2.9). Therefore since W starts and ends with the same P_j- or R_j-letter, the free computation $W \bullet h$ corresponds to a spiral whose inner contour has the same label as the inner contour of Γ and the outer contour has the same label mod $\mathcal{G}$ as the outer contour of Γ. Therefore the inner parts of non-P_1z- and non zR_1-sectors of $W \circ h$ coincide with the corresponding subwords of W'. We would be able to apply the same argument as in Case 1.1 to the spiral corresponding to the free computation $W \bullet h$ if we bound the sizes of P_1z- and zR_1-sectors of the word $W \circ h$.

Let us prove that these sizes are recursively bounded. We can certainly assume by Remark 8.1, that the base $W^{\pm 1}$ starts and ends on P_1 or R_1. If the base of $W^{\pm 1}$ contains both P_1 and R_1, we can assume that it starts with P_1.

By Lemma 2.2 there are the following possibilities for the base of $W^{\pm 1}$:

- the base of $W^{\pm 1}$ has a prefix $P_1 R_1 K_2$ and suffix $K_1 L_1 P_1$;

- the base of $W^{\pm 1}$ has a prefix $P_1 P_1^{-1}$ and suffix $K_1 L_1 P_1$;

- the base of $W^{\pm 1}$ has a prefix $R_1^{-1} R_1 K_2$.

These three cases are completely similar, so let us consider only the first one. Then $W^{\pm 1}$ has the form $P_1 V_1 R_1 V_2 K_2 ... K_1 V_3 L_1 V_4 P_1$ where V_1 (resp V_2, V_3, V_4) is a word in $\mathcal{A}(P_1)$ (resp. $\mathcal{A}(R_1), \mathcal{A}(K_1), \mathcal{A}(L_1)$). By Lemma 2.6 there exist words $u(L_1, h_0)$, $v(L_1, h_0)$, $u(P_1, h_0)$, $v(P_1, h_0)$, $u(R_1, h_0)$, $v(R_1, h_0)$ such that

$$\begin{aligned}&\beta(u(L_1, h_0)) \equiv \beta(v(L_1, h_0))^{-1}, \\ &\beta(u(P_1, h_0)) \equiv \beta(v(P_1, h_0))^{-1}, \beta(u(R_1, h_0)) \equiv \beta(v(R_1, h_0))^{-1}\end{aligned} \quad (8.1)$$

and for every $t = 1, ..., s$

$$W \circ h_0^t = P_1 u(P_1, h_0)^t V_1 v(R_1, h_0)^t R_1 u(R_1, h_0)^t V_2 K_2^{-1} \\ ... K_1 V_3 v(L_1, h_0)^t L_1 u(L_1, h_0)^t V_4 v(P_1, h_0)^t P_1.$$

Since we assumed that $s > |q| + |p|$, $|V_3| < |q|$ and $V_3 v(L_1, h_0)^t \leq |p|$, we must have $v(L_1, h_0) = 1$ in the free group. Hence by (8.1) $u(L_1, h_0) = 1$ in the free group. Since $|V_4| < |q|$ and $|V_4 v(P_1, h_0)^s| < |p|$, we have that $v(P_1, h_0) = 1$ in the free group. Then $u(P_1, h_0)$ is freely equal to 1. Similarly, considering the words V_2 and $v(R_1, h_0)^s V_2$, we conclude that $v(R_1, h_0) = 1$ in the free group. Hence $u(R_1, h_0)$ is freely trivial. But then $|u(P_1, h_0)^s V_1 v(R_1, h_0)^s| = |V_1| < |q|$, and we have a (linear) bound of the size of the $P_1 R_1$-sector of the word $W \circ h$ as desired.

Case 2. Suppose now that there are neither P_j- nor R_j-letters in the base of Δ.

Then the trapezium Δ has no $\mathcal{G}$-cells by Lemma 6.5. Again, it suffices to give a recursive upper bound for the length of the loops $q(l)$, because this would give a recursive estimate for the number of their labels. As in Case 1, we can bound the sizes of the paths $p(l)$. Since the history of Δ is periodic and the lengths of the words $\phi(p(l))$ (whose (k,a)-projections are the intermediate words, without the last letter, of the corresponding computation) are bounded, the words $\phi(p(l))$ form a periodic sequence whose period n is recursively bounded in terms of $|q|+|p|$.

Thus it remains to obtain an upper bound for the lengths of the paths $t(l)$. Since the number of θ-edges in $t(l)$ is at most $\min(|q|,|p|)$, it suffices to obtain an upper estimate for the length of each of the x-subpaths t_{lj} of $t(l)$ connecting neighbor θ-bands of Δ.

The θ-bands in Γ subdivide this diagram into several (at most $|q|$) subdiagrams Γ_i containing no θ-cells. The path t_{lj} is a side of one of the k-bands in one of the subdiagrams Γ_i. Thus it is to show that k- and a-bands in Γ_i have bounded lengths. Notice that the boundary of the van Kampen diagram Γ_i has the form $q^1 q^2 q^3 q^4$ where $\phi(q^1)$ and $\phi(q^3)$ are subwords in $\mathcal{X}$ of the boundary labels of Δ and $\phi(q^2), \phi(q^4)$ are subwords of the top and the bottom labels of θ-bands in Γ.

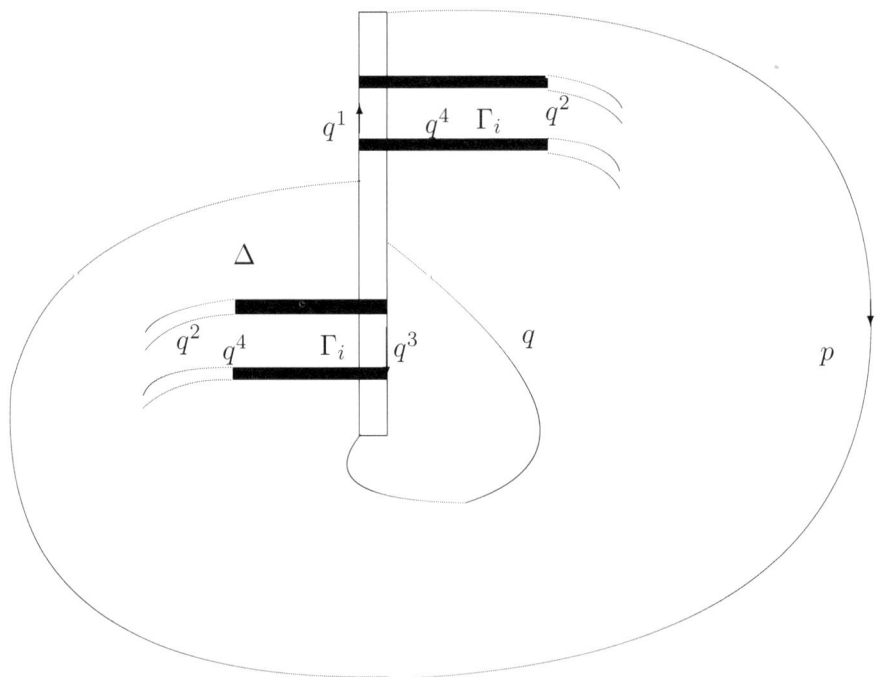

Fig. 23.

The words $\phi(q^2)$, $\phi(q^4)$ are periodic with the length of the period bounded in terms of n. Hence Γ_i can be cut by its k- or a-bands into subdiagrams $\Psi_1, \Psi_2, \ldots \Psi_r$ with contours $q_i^1 q_i^2 q_i^3 q_i^4$ where q_i^1, q_i^3 are words in $\mathcal{X}$, q_i^2 (or q_i^4) have the same labels of length n for all i, with the possible exception for q_r^2, q_r^4 whose lengths are recursively bounded in terms of n. By Lemma 8.2, the subdiagram Ψ_r has bounded number of cells and can be ignored. Also Lemma 8.2 implies that it suffices to bound $|q_i^1|$ for all i in terms of $|q_1^1|, |q_{r-1}^1|$.

Let $w_i \equiv \phi(q_i^1)$. By Lemma 8.2 for Ψ_i, we have that w_{i+1} is freely equal to $uw_i'v$ where the fractional words u, v depend on $\phi(q_i^2), \phi(q_i^4)$, and therefore, do not depend on i, and so they belong to some finite effectively determined set of words; w_i' results from w_i after a substitution of the form $x \mapsto (x')^{4^d}$ for a recursively bounded d.

If $d = 0$, then we have, by the obvious induction, that w_i is freely equal to a copy of $u^{i-1} w_1 v^{i-1}$. Then, by Lemma 8.1, $|w_i|$ is effectively bounded by $|w_1| + |w_{z-1}|$, as desired. Hence we may assume that $d \geq 1$. (We would interchange the top and the bottom of the spiral if $d \leq -1$.)

Since $d \geq 1$, $|w_{i+1}| \geq 4|w_i| - C$ where $C = |u| + |v|$. Then either $|w_i| \leq C$ for every i, or, if $|w_i| > C$ for some i, the series $|w_i|, |w_{i+1}|, \ldots$ monotonically increases, and so $|w_i| \leq |w_{z-1}|$.

In any case the lengths of the words w_i are recursively bounded in terms of $|q| + |p|$ and the lemma is proved. $\square$

Lemma 8.7. *Let a minimal diagram Δ be a quasispiral having no k-cells. Then the length of a shortest a-band of Δ, which connects p and q is bounded by a recursive function of $|q| + |p|$, where q, p are the inner and the outer contours of Δ.*

Proof. One can define turns of the quasispiral as for spirals (in the beginning of the proof of Lemma 8.6 (but one should choose a maximal a-band instead of $\mathcal{B}_k$). All the turns have equal boundary labels since every maximal a-band has one common cell with every turn. (Recall that there are no terminal cells for a-bands in Δ, because the base of Δ is empty.) Then one can apply Lemma 8.2 exactly as in Case 2 of the proof of Lemma 8.6 to complete the proof. $\square$

9 Rolls

A reduced annular diagram Δ over the group $\mathcal{H}_1$ is said to be a *roll* if its inner contour q and its outer contour p are minimal boundaries and have no k-letters (see Figure 24).

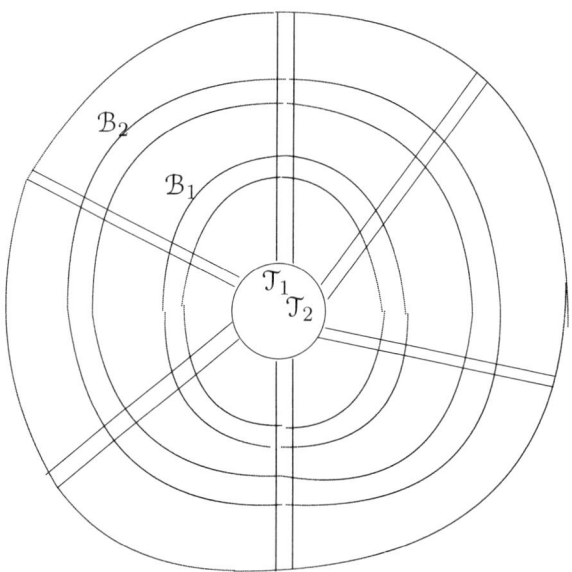

Figure 24.

It follows that every maximal θ-band of a roll Δ connects its inner boundary q and outer boundary p. The cyclic *history* of a roll is the projection of $\phi(p)$ on $S \cup \bar{S}$ (it is equal to the projection of $\phi(q)$ on $S \cup \bar{S}$). Clearly (by Lemma 3.11) maximal k-bands of a roll Δ form concentric annuli surrounding the hole of the diagram. It follows from the definition that arbitrary annular subdiagram of a roll bounded by some sides of k- or a-annuli is a roll if it contains a (θ, k)-cell. In any case it is a roll if it is bounded by sides of two a-annuli, since there are no a-edges on these sides.

We count the k-annuli of Δ from q to p: $\mathcal{B}_1, ..., \mathcal{B}_r$. The bases of $\mathcal{B}_1, ..., \mathcal{B}_r$ form the *base* of the roll Δ.

Our goal is to replace every roll by a roll with the same boundary labels and recursively bounded number of cells (in terms of $|p| + |q|$).

9.1 Rolls without a base

Let $z \in \tilde{\mathcal{K}}$. We say that an x-edge (a-edge, θ-edge) e is of *type z* if the label of $e^{\pm 1}$ belongs to the set $\mathcal{X}(z)$ (to $\mathcal{A}(z) \cup \bar{\mathcal{A}}(z)$, to $\Theta(z) \cup \bar{\Theta}(z)$).

Lemma 9.1. *Let Δ be a roll with empty base. Suppose that either*
(1) the contours p and q are simple loops without common vertices or
(2) all edges in $p \cup q$ are of the same type z for some $z \in \tilde{\mathcal{K}}$.
Then the labels of all edges in Δ are of the same type.

Proof. It is clear from the list of relations of $\mathcal{H}_1$, that all edges of a non-k-cell have the same type. We say that two cells of Δ are neighbor if they have a common edge. It follows that the transitive closure of this relation is an equivalence, and all edges of the cells of an equivalence class Γ are of the same type.

Let $\Gamma_1, \ldots, \Gamma_s$ be all the classes of Δ. If $s > 1$, then the deletion of finitely many vertices from $\cup_{i=1}^{s}\Gamma_i$ makes Δ disconnected, a contradiction in case (1). Hence $s = 1$ in this case, and the statement is true because every edge of $p \cup q$ must belong to a cell of Γ_1. In case (2) the statement follows from the observation that any boundary edge of each of the components $\Gamma_1, \ldots, \Gamma_s$ must lie on $p \cup q$.

□

Lemma 9.2. *There is a recursive function f with the following property.*

Let Δ be a roll with an empty base. Then there exists a roll Δ' with the same boundary labels as Δ, which has a path of length at most $f(|p|+|q|)$, connecting the components p and q of the boundary of Δ'.

Proof. We can assume that Δ contains θ-edges since otherwise one can refer to Lemma 4.4. By Lemma 5.5, we can assume that Δ satisfies properties (R1) - (R4).

Case 1: Assume that there is at least one $\mathcal{A}(P_1) \cup \bar{\mathcal{A}}(P_1)$-edge in Δ. By Lemmas 9.1 and 5.1, we can assume that all a-edges in Δ belong to $\mathcal{A}(P_1) \cup \bar{\mathcal{A}}(P_1)$, all θ-edges belong to $\Theta(P_1) \cup \bar{\Theta}(P_1)$. Since the set $\mathcal{X}(P_1)$ does not exist, Δ does not have x-edges.

Thus all cells in Δ correspond to the commutativity relations of the form (2.7) and $\mathcal{G}$-relations. Therefore Δ is a diagram over the direct product of the free group generated by Θ and the free product of the copy of $\mathcal{G}$ generated by $\mathcal{A}(P_1)$ and the free group with basis $\{a_{m+1}, \ldots, a_{\bar{m}}\}$. Since the conjugacy problem is decidable in both factors of this direct product, we can replace Δ by a diagram with the same boundary label which has a path, connecting the inner and outer contours, of length bounded by a recursive function in $|p| + |q|$, as required.

Case 2. Suppose that Δ has no $\mathcal{A}(P_1)$-edges.

Again by Lemma 9.1, for some $z \in \tilde{\mathcal{K}}$, $z \neq P_1$, all edges from Δ are of type z.

Case 2.1. First we consider the case where p and q have no a-edges. By Lemma 3.11, Δ is a union of concentric a-annuli $\mathcal{C}^1, \mathcal{C}^2, \ldots$. If we find a recursive upper bound for the lengths of the a-annuli, we recursively bound their number because if the sides of two such annuli have equal labels, we can remove the annuli and all the cells between them to obtain a smaller roll. Since all the a-annuli have equal numbers of θ-cells (each of them intersects each of the θ-bands only once), it suffices to bound the lengths of the subbands of $\mathcal{C}^i$ between any two neighbor maximal θ-bands $\mathcal{T}$ and $\mathcal{T}'$. Let Γ be the subdiagram situated between $\mathcal{T}$ and $\mathcal{T}'$. Clearly all cells in Γ correspond to the relations (2.8).

We can assume that $z \neq P_j$ for any j since there are no auxiliary $(\mathcal{A}(P_j), x)$-relations (2.8). Hence $z \in \{K_j, L_j, R_j\}$ for some j.

We consider only the first of these cases because all other cases are similar. Thus assume that $z = K_j$ for some j.

The bands $\mathcal{T}$ and $\mathcal{T}'$ are reduced, and so Γ has boundary label of the form $U_1 V_1 U_2 V_2$ where V_1 and V_2 are labels of sides of $\mathcal{T}$ and $\mathcal{T}'$, and U_1 and U_2 are words in $\mathcal{X}$. Clearly the a-projections of V_1 and V_2^{-1} are graphically the same since every a-band starting on $\mathcal{T}$ ends on $\mathcal{T}'$ and vice versa. Let $V = (V_1)_a$.

Lemma 6.11 can be applied to a minimal subdiagram Γ' of Γ, containing two neighbor a-bands. It implies that the word V is a product $V'(V'')^{-1}$ for some positive words V', V''. Now Lemma 6.3 implies that all conditions of Lemma 4.5 hold for the union of Γ' and the corresponding portions of the θ-bands $\mathcal{T}, \mathcal{T}'$. Lemma 4.5 gives the desired recursive bound for the lengths of a-bands in Γ in terms $|U_1|, |U_2|$.

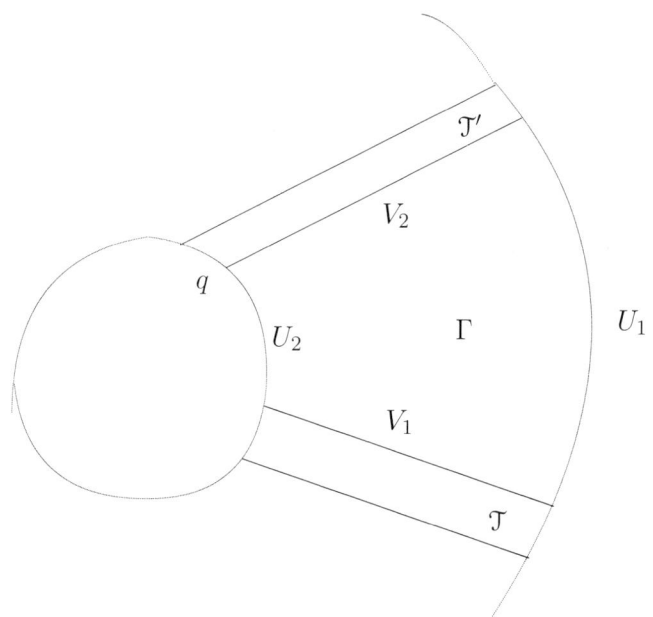

Fig. 25.

Case 2.2 Now let us consider the case when $\partial(\Delta)$ contains a-edges. In this case, there is an a-band $\mathcal{C}$ starting on p and ending on q since Δ satisfies (R2). So we can apply Lemma 8.4 and represent Δ as a union of a quasispiral, and two subdiagrams with recursively bounded number of cells. Then the sub-quasispiral of Δ will have a recursively bounded perimeter, and we can apply Lemma 8.7. □

Lemma 9.3. *Let Δ be a roll with the empty base, satisfying (R3), which has no $\mathcal{G}$-cells. Let p and q be the outer and inner components of $\partial(\Delta)$*

and $\phi(p)_a$ freely equal to the empty word. Then either there is an a-annulus between p and q surrounding the hole of Δ, or the lengths of the θ-bands of Δ are recursively bounded in terms of the number of a-letters in $\phi(p)$ and $\phi(q)$.

Proof. By Lemma 5.5, we can assume that no a-bands in Δ start and end on p (resp. q). Therefore we can assume that all a-bands in Δ connect p and q (otherwise it would contain an a-annulus).

Since the a-projection of $\phi(p)$ is freely trivial, there is a subpath p' of p which starts with an edge e, ends with e' and $\phi(e) = \phi(e')^{-1} \in \mathcal{A} \cup \bar{\mathcal{A}}$, and there are no a-edges in p' between e and e'. Consider the a-bands $\mathcal{C}$ and $\mathcal{C}'$ starting on e and e' respectively. Suppose that both bands intersect the same θ-band $\mathcal{T}$. Then the intersection cells $\mathcal{C} \cap \mathcal{T}$ and $\mathcal{C}' \cap \mathcal{T}$ must have a common θ-edge, and their a-edges must have mutually inverse labels. Thus these cells cancel, a contradiction. Hence, say, $\mathcal{C}$ does not intersect all maximal θ-bands in Δ. This and the Jordan lemma implies that none of the a-bands in Δ intersects any of the maximal θ-bands there more than ones. Therefore by Lemma 3.11 each a-band intersects each θ-band in Δ at most once. Hence the length of each θ-band in Δ does not exceed the number of a-edges on p. $\square$

9.2 Rolls with L-annuli

Lemma 9.4. *Let a reduced annular diagram Δ be an L_j-annulus with an outer boundary p and inner boundary q, whose cyclic history is a word of the form $\tau\tau^{-1}\tau\tau^{-1}\ldots$ for some $\tau \in \mathcal{S}^+(2) \cup \mathcal{S}^+(4)$. Assume that all x-edges of p are labelled by the same letter $x = x(a,\tau)$, $a \in \mathcal{A}(L_j)$. Then there exists an annular diagram Δ' which contains no k-cells with the outer path p' and inner path q', where $\phi(p') \equiv \phi(p)$ and $\phi(q')$ can be obtained from $\phi(q)$ by replacing every a-, x-, and θ-letter by its brother in $\mathcal{A}(L_j) \cup \mathcal{X}(L_j) \cup \Theta(L_j)$. Similarly there also exists an annular diagram Δ'' with outer contour p'' and inner contour q'', without k-cells where all a-, x-, and θ-labels of the boundary belong to $\mathcal{A}(\overleftarrow{L_j}) \cup \mathcal{X}(\overleftarrow{L_j}) \cup \Theta(\overleftarrow{L_j})$, $\phi(q'') \equiv \phi(q)$, $\phi(p'')$ is a copy of $\phi(p)$.*

Proof. Assume that $\tau \in \mathcal{S}^+(2)$ (the other case is similar). Consider an (L_j, θ)-cell Π of Δ with boundary label of the form

$$\theta_1^{-1} L_j \theta_2 x^{-1} a L_j^{-1} (a')^{-1} x'$$

(see relations (2.6)) where $a \in \mathcal{A}(L_j)$, $a' \in \mathcal{A}(\overleftarrow{L_j})$, $x \in \mathcal{X}(L_j)$, $x' \in \mathcal{X}(\overleftarrow{L_j})$. We substitute the occurrence L_j by $x'a'$, the occurrence L_j^{-1} by $(a')^{-1}x'$, the letter θ_2 by its copy θ_1, the letter a by a' and x by x'. Then we obtain a relation of the form (2.7). We can substitute the k-cell Π by the new cell Π' without k-edges.

Then, for every (L_j, x)-cell π (see relations (2.9)), situated between (L_j, θ)-cells Π_1 and Π_2, we change the letters L_j in its label by $x'a'$ if the subtrapezium, which includes Π_1 and Π_2, has history $\tau^{-1}\tau$, or by $(x')^{-1}a'$ if the history is $\tau\tau^{-1}$. The letter x in the boundary label of π will be replaced by x'. We obtain the boundary label of the form $(x')^4(x')^{\pm 1}a'(x')^{-1}(a')^{-1}x^{\mp 1}$ which is equal to 1 modulo relations (2.8).

Since the cells of Δ were glued in the L_j-band along L_j-edges, the substitutions we have just described provide us with the desired diagram Δ'. □

Lemma 9.5. *Let a roll Δ with base $L_j L_j^{-1}$ or $L_j^{-1} L_j$ be bounded by two L_j-annuli $\mathcal{B}_1$ and $\mathcal{B}_2$. Assume that the non-empty history of the roll is $\tau\tau^{-1}\tau\tau^{-1}\ldots$ for some $\tau \in \mathcal{S}^+(2) \cup \mathcal{S}^+(4)$. Suppose that all x-edges of the inner boundary of $\mathcal{B}_1$ have the same labels and all x-edges of the outer boundary of $\mathcal{B}_2$ have the same labels $x(a, \tau)$ (for some a). Then there exists a roll Δ' which has no k-cells and the same boundary labels as Δ.*

Proof. By Lemma 3.11, every θ-band of Δ crosses each of $\mathcal{B}_1, \mathcal{B}_2$ only once. Therefore one can apply Lemma 6.5 (part 2) to conclude that Δ contains no $\mathcal{G}$-cells. Besides, by Lemma 9.1 (assumption 2 of Lemma 9.1 holds), the subroll with trivial base bounded by $\mathcal{B}_1$ and $\mathcal{B}_2$, consists of $\mathcal{A}(L_j)$-cells or it consists of $\mathcal{A}(\overleftarrow{L}_j)$-cells. Notice that for every $\mathcal{A}(L_j)$-cell, can be constructed its $\mathcal{A}(\overleftarrow{L}_j)$-copy, and vice versa (see relations (2.7), (2.8) and the definition of the mapping α_τ). Then one can change all these cells by their $\mathcal{A}(\overleftarrow{L}_j)$-copies or, respectively, by their $\mathcal{A}(L_j)$-copies and simultaneously transform the bands $\mathcal{B}_1$ and $\mathcal{B}_2$ according to Lemma 9.4. Reducing the resulting diagram, we obtain the desired roll Δ'. □

Lemma 9.6. *Let Δ be a roll with a non-empty history h which is a word in $\mathcal{S}(2) \cup \mathcal{S}(4)$. Assume that the base of Δ is $L_j^{-1} L_j$ for some j, and Δ is bounded by two L_j-annuli $\mathcal{B}_1$ and $\mathcal{B}_2$. Then h is a word of the form $\tau\tau^{-1}\tau\ldots$ for some $\tau \in \mathcal{S}$, all x-edges in the boundaries p or q are labelled by the same letter $x = x(a, \tau)^{\pm 1}$, and all x-edges not in $p \cup q$ are also labelled by the same letter $x' = x(a', \tau)^{\pm 1}$.*

Proof. Assume that the history h, as a cyclic word, has two neighbor letters τ, τ' which are not mutually inverse. Consider the small subtrapezium Γ of Δ with history $\tau\tau'$ and a two-letter base. Since the history of Γ is reduced, by Lemma 6.12, the projection of the label of the bottom of the trapezium Γ on $\mathcal{A} \cup \mathcal{K}$ is an admissible word for $\mathcal{S}$. But this contradicts Lemma 2.2 (an admissible word for $\mathcal{S}$ cannot have base $L_j^{-1} L_j$). Hence h does not have a reduced subword of length 2, whence it has the form $\tau\tau^{-1}\tau\ldots$ for some τ.

Let Δ' be the roll with empty base obtained by removing $\mathcal{B}_1, \mathcal{B}_2$. Since the history of the roll has the form $\tau\tau^{-1}\ldots$, there exists i such that all a-edges on the boundary of the roll have labels $a_i(z)^{\pm 1}$ for some i where $z = \overleftarrow{L}_j$. By Lemma 9.1(2), all edges of Δ' are of type $\overleftarrow{L}_j$.

Let Γ be a small subtrapezium of Δ bounded by two maximal Θ-bands $\mathcal{T}_1, \mathcal{T}_2$ of Δ starting on q and ending on p. Denote $\pi_1 = \mathcal{T}_1 \cap \mathcal{B}_1$ and $\pi_2 = \mathcal{T}_2 \cup \mathcal{B}_1$.

We assume that $\mathcal{T}_2$ follows $\mathcal{T}_1$ in the clockwise order. Let
$$V \equiv \phi(\mathbf{bot}(\mathcal{T}_1)), V' \equiv \phi(\mathbf{top}(\mathcal{T}_2)).$$
Since every maximal a-band of Γ starting on $\mathbf{top}(\mathcal{T}_1)$ ends on $\mathbf{bot}(\mathcal{T}_2)$, the a-projections V_a and V'_a are identical and each of these words contains a letter $b^{\pm 1} = \in \mathcal{A}(\overleftarrow{L}_j)^{\pm 1}$.

Suppose that the first a-letter of the word V_a is negative and the last one is positive. Then V_a has a subword of the form $a_1^{-1} a_2$ which is ruled out by Lemma 6.11.

Therefore either the first letter of V_a is positive or the last letter is negative. These two cases are similar (one can be obtained from the other one by interchanging of p and q).

Thus we may assume that the first letter of V_a is positive. Hence by Lemma 6.11, $V = V_1 V_2$ where $(V_1)_a$ is a non-empty positive word and $(V_2)_a$ is a negative word.

Again, we consider the first $a_s(\overleftarrow{L}_j)$-band $\mathcal{C}$ in Γ, counting from q, connecting $\mathbf{bot}(\mathcal{T}_1)$ and $\mathbf{top}(\mathcal{T}_2)$. Let Ψ be the diagram containing $\mathcal{C}$ and bounded by a part of $\mathbf{bot}(\mathcal{T}_1)$, $\mathbf{bot}(\mathcal{C})$, a part of $\mathbf{top}(\mathcal{T}_2)$ and a part of the outer contour of $\mathcal{B}_1$.

The boundary label of Ψ is of the form $W \equiv U_1 a_s(\overleftarrow{L}_j)^{-1} x^{\pm 1} U_2 x^{\mp 1} a_s(\overleftarrow{L}_j)$, where U_1 and U_2 are some words in $\mathcal{X}$, U_1 is the top label of the $a_s(\overleftarrow{L}_j)$-band $\mathcal{C}$ and U_2 is read on the outer contour of $\mathcal{B}_1$ between two L_j-cells. Hence the word U_2 is freely equal to a product of fourth powers of x-letters labelling x-edges of the outer side of $\mathcal{B}_1$ (by relations (2.9)). This word cannot be empty, because otherwise the two L_j-cells π_1, π_2 would form a reducible pair. The letter x labels either an x-edge of the L_j-cell π_1 (where $\mathcal{T}_1$ starts) or an x-edge of an auxiliary θ-cell π of $\mathcal{T}_1$ (that could happen either if the cell π_1 has no a-edges on its top side, or if the first auxiliary cell in $\mathcal{T}_1$ shares its x-edge with π_1). We shall refer to these possibilities as **Cases 1** and **2**.

Notice that the letter x labels an x-edge of a θ-cell. Therefore it corresponds to the a-letter appearing as a label of an a-edge of this cell and the S-rule τ. We shall prove that all x-edges in Δ' are labelled by $x^{\pm 1}$, and that all a-edges on the sides of the θ-bands of Δ' are labelled by $a_s(\overleftarrow{L}_j)^{\pm 1}$. Then we shall prove that $s = i$. This would imply that all x-edges on the boundary of Δ have the same label.

The word $x^{\pm 1} U_2 x^{\mp 1}$ is freely inverse of the bottom label U'_1 of $\mathcal{C}$. Therefore, by Lemma 6.10(2), $U_2 = x^{4d}$ for some $d \neq 0$. In particular, the letter $a_s(\overleftarrow{L}_j)$ is determined by the x-letter x' such that x is a copy of x' and $(x')^d$ is read on p between π_1 and π_2. Thus, $U_1 = x^{-d}$.

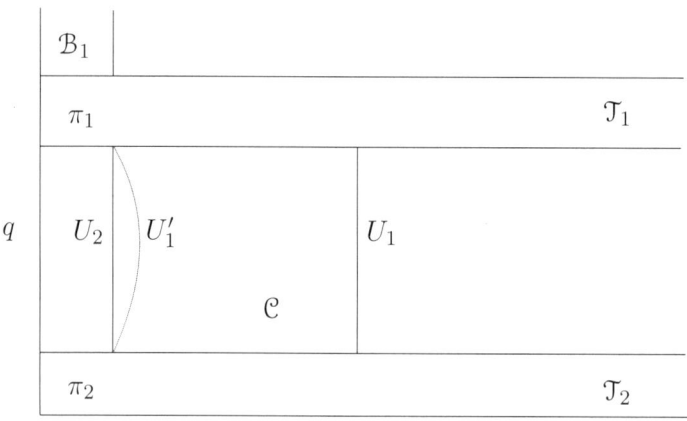

Fig. 26.

If $|(V_1)_a| \geq 2$, one can almost repeat the argument (but applying now part (3) of Lemma 6.10 instead of part (2)) and conclude that the next to the first a-letter $a_{s'}(\overleftarrow{L}_j)$ of $(V'')_a$ is determined by x, and so $s' = s$. Hence $(V_1)_a$ is a power of $a_s(\overleftarrow{L}_j)$ and all x-edges of the a-bands starting on the portion of $\mathbf{bot}(\mathcal{T}_1)$ labelled by V_1 have the same label x. Similarly, if $(V_2)_a$ is non-empty, it must be a power of some $a_{s'}(\overleftarrow{L}_j)$, and all x-edges on a-bands starting on the part of $\mathbf{bot}(\mathcal{T}_1)$ labelled by V_2 have the same label x''. The neighbor $a_s(\overleftarrow{L}_j)$- and $a_{s'}(\overleftarrow{L}_j)$-bands, connecting $\mathbf{bot}(\mathcal{T}_1)$ and $\mathbf{top}(\mathcal{T}_2)$, have a common boundary sections labelled by non-empty $x^{t_1} \equiv (x'')^{t_2}$, because otherwise a cell of $\mathcal{T}_1$ and a cell of $\mathcal{T}_2$ would form a reducible pair. We conclude that $x \equiv x''$ and hence $s = s'$. Therefore V_2 is empty. Similarly $\phi(\mathbf{top}(\mathcal{T}_2))_a$ is a power of $a_s(\overleftarrow{L}_j)$.

Considering the next (in the clockwise order) small subtrapezium Γ' of Δ', the small subtrapezium after that, and so on, we conclude that all a-edges on the (reduced) top and bottom sides of all θ-bands in Δ' have the same label $a_s(\overleftarrow{L}_j)$ and all x-edges in all small subtrapezia of Δ' are labelled by x.

It remains to show that $s = i$. Indeed, if Case 1 holds, that is if π_1 does not share its x-edge with the first auxiliary cell π of $\mathcal{T}_1$, then the first a-edge of $\mathbf{bot}(\mathcal{T}_1)$ is an edge of π_1, and so $s = i$. Suppose that Case 2 holds. Then the a-edges of π are labelled by $a_i(\overleftarrow{L}_j)^{\pm 1}$.

The bottom side of π_1 (considered as a θ-band) consists of one L_j-edge (see relations (2.6)), so the bottom side of π does not have common edges with the bottom side of π_1. Therefore the a-edge of the bottom side of π appears on the (reduced) bottom path of $\mathcal{T}_1$. Therefore, again, $s = i$. This completes the proof of the lemma. □

Lemma 9.7. *Let Δ be a roll with non-empty base whose history is a word over $\mathcal{S}(2) \cup \mathcal{S}(4) \cup \bar{\mathcal{S}}$. Then either all rules in the history belong to $\bar{\mathcal{S}}$ or all of these rules are from $\mathcal{S}(2) \cup \mathcal{S}(4)$.*

Proof. Indeed, consider a k-annulus $\mathcal{B}$ of Δ. If the history of Δ contains rules both from $\bar{\mathcal{S}}$ and from $\mathcal{S}(2) \cup \mathcal{S}(4)$ then there exist two neighbor cells in $\mathcal{B}$, one corresponding to a rule in $\bar{\mathcal{S}}$ and the other corresponding to a rule from $\mathcal{S}(2) \cup \mathcal{S}(4)$. But this is impossible since rules from $\mathcal{S}(2) \cup \mathcal{S}(4)$ do not change the Ω-coordinate of the base letters, and letters in $\mathcal{K} \cap \bar{\mathcal{K}}$ have Ω-coordinates 1. $\square$

Lemma 9.8. *Let Δ be a roll with base $L_j^{-1} L_j$ for some j. Assume that the non-empty history h of Δ is a word in $\mathcal{S}(2) \cup \mathcal{S}(4) \cup \bar{\mathcal{S}}$. Then there exists a roll Δ' with an empty base, having the same boundary labels as Δ.*

Proof. One may assume that Δ is bounded by two L_j-annuli $\mathcal{B}_1$ and $\mathcal{B}_2$. By Lemma 9.7, there are two possibilities: either all rules in the history of Δ are from $\bar{\mathcal{S}}$ or all of them are from $\mathcal{S}(2) \cup \mathcal{S}(4)$.

In the first case the inner side of $\mathcal{B}_1$ has the same label as the outer side of $\mathcal{B}_2$ since there exists no $(\bar{a}, x)$-cells. Hence the roll Δ can be replaced by an empty roll.

In the second case, one can apply Lemma 9.6 and then Lemma 9.5. $\square$

Lemma 9.9. *Let Δ be a roll. Assume that the history h of Δ is non-empty and the base of Δ is either (a) equal to $L_j^{-1} L_j$ or (b) equal to $L_j L_j^{-1}$ or (c) equal to L_j or for some j. Then there exists a roll Δ' with the same boundary labels as Δ, whose base is obtained from the base of Δ by removing some (or none) letters, and and the number and perimeters of cells are recursively bounded in terms of lengths of the boundary components $|p|, |q|$ (as usual, p is the outer and q is the inner boundary components).*

Proof. 1. By Lemma 5.5 we can assume that the roll Δ satisfies the conditions (R1) and (R2) of that lemma. We assume that Δ has the lowest possible number of L_j-annuli and the lowest type among all rolls satisfying the conditions of the lemma, plus conditions (R1), (R2), and having the same boundary labels as Δ.

If we remove the L_j-bands from Δ, we get two or three rolls with empty bases. Two of these rolls, say, Δ_1 and Δ_2 are bounded by an L_j-annulus and q and an L_j-annulus and p, respectively, and the third roll Δ_3 (if exists) is bounded by two L_j-annuli.

Notice that it is enough to be able to replace Δ by a roll with the same boundary labels and recursively bounded lengths of L_j-bands. Indeed in that case we would be able to apply Lemma 9.2 to the subrolls $\Delta_1, \Delta_2, \Delta_3$.

2. Fix a big enough recursive function $f(x)$; it will be clear later how big $f(x)$ should be: $f(x)$ has to satisfy certain (finite) number of inequalities of the form $f(x) > g(x)$ where $g(x)$ is a recursive function.

3. Suppose that one can connect the boundary components of Δ_i, $i = 1, 2$, by a path of length $\leq f(|p| + |q|)$. Then by Lemma 5.3, the number of cells in Δ_i is recursively bounded in terms of $|p|, |q|$. Thus we can remove one of the subrolls Δ_1 or Δ_2 and one of the L_j-annuli from Δ and reduce the problem to the same problem about a roll with fewer L_j-annuli.

Thus we can assume that there are no paths of length $\leq f(|p| + |q|)$ connecting the boundary components of Δ_i, $i = 1, 2$ (if Δ_i does exist). By Lemma 5.1, we may also assume that p and q are simple loops. By Lemma 9.1, we conclude that for some $z \in \{L_j, \overleftarrow{L_j}\}$, all edges of the diagram Δ_i are of the same type z. In particular, Δ does not contain $\mathcal{G}$-cells.

4. As a consequence of 3, we can deduce that the lengths of maximal θ-bands in Δ_1 and Δ_2 are bigger than 1 (since $f(x) > 1$). Hence for every τ in the history of Δ there exist relations of the form (2.7) containing a letter from $\Theta(\overleftarrow{L_j}) \cup \bar{\Theta}(\overleftarrow{L_j})$. This implies that τ does not lock (zL_j)-sectors and (L_jz)-sectors. So every rule τ in the history of Δ belongs to $\mathcal{S}(2) \cup \mathcal{S}(4) \cup \bar{\mathcal{S}}(2) \cup \bar{\mathcal{S}}(4)$. Hence if one of the rules in that history belongs to $\bar{\mathcal{S}}$, all of them belong to $\bar{\mathcal{S}}$, which implies that Δ_i ($i = 1, 2$) does not contain x-edges at all (there are no letters in $\mathcal{X}$ that correspond to $\bar{\mathcal{A}}$-letters). This, in turn, would imply that the lengths of the L_j-annuli in Δ equal the number of maximal θ-bands starting on p and ending on q, hence the lengths of L_j-annuli in Δ are bounded in terms of $|p|, |q|$, as desired. Hence we can assume that all rules τ in the history of Δ belong to $\mathcal{S}(2)$ or $\mathcal{S}(4)$.

5. Now if the base of Δ is $L_j^{-1} L_j$ (as in case (a) of the lemma), we can apply Lemma 9.8 (to obtain a roll with the same boundary labels as Δ but with empty base) and then use Lemma 9.2.

6. Thus suppose that the base of Δ is L_j or $L_j L_j^{-1}$. Then $z_1 = K_s^{\pm 1}$ for some s.

Consider two arbitrary consecutive (in the clockwise order) maximal θ-bands $\mathcal{T}_1$ and $\mathcal{T}_2$ in $\Delta_1 \cup \mathcal{B}_1$ starting on the inner contour q. Let Γ be the subdiagram of $\Delta_1 \cup \mathcal{B}_1$ bounded by $\mathbf{bot}(\mathcal{T}_1)$ and $\mathbf{top}(\mathcal{T}_2)$. Fewer than $|q|$ of the maximal a-bands in Γ that start on $\mathbf{top}(\mathcal{T}_1)$ end on q. The total number of cells in these bands is recursively bounded (we can remove these bands one by one starting with the band which shares one of its sides with q). Thus we can remove these cells from Δ. Similar operation can be performed on all other subdiagrams of Δ_1 situated between two consecutive θ-bands, since the number of such diagrams is at most $|q|$. Thus we can assume that every maximal a-band in Γ starting on $\mathbf{bot}(\mathcal{T}_1)$ ends on $\mathbf{top}(\mathcal{T}_2)$. Let $\mathcal{C}$ be the first of these a-bands (counting from q to $\mathcal{B}_1$). Then the bottom side of $\mathcal{C}$ is a part of q.

Note that the rules of $\mathcal{S}(2) \cup \mathcal{S}(4)$ have the form $[..., K_s \to K_s, ...]$. Therefore the corresponding relations 2.6 are of the form

$$\theta(\tau,(K_s)_-)^{-1}K_s(r,\omega)\theta(\tau,K_s) = K_s(r,\omega)$$

where $\omega \in \{2,4\}$, $r \in \bar{\mathcal{E}}$. Using cells corresponding to these relations and relations of the form (2.9), we can build a K_s-band $\mathcal{B}$ whose top side has the same label as the bottom side of $\mathcal{C}$. Then we can attach $\mathcal{B}$ to the bottom side of $\mathcal{C}$ and obtain a small trapezium $\tilde{\Gamma} = \Gamma \cup \mathcal{B}$.

7. Suppose that the history of $\tilde{\Gamma}$ (= the history of Γ) is reduced. Then by Lemma 6.12, the labels of the top and bottom paths of $\tilde{\Gamma}$ are admissible words. In particular it means that the a-projection of the word V (resp. V') written on the bottom (resp. top) path of the trapezium $\tilde{\Gamma}$ is positive. But then Lemma 4.5 (iii) implies that the length of $\mathcal{T}_1$ does not exceed $2|\mathcal{C}|+2$. This contradicts our assumption that it is greater than $f(|p|+|q|)$.

Hence the history of Γ is not reduced. Since Γ was chosen arbitrarily, we deduce that the history of Δ has the form $\tau\tau^{-1}...$ for some $\tau \in \mathcal{S}(2) \cup \mathcal{S}(4)$.

8. Let $V = \phi(\mathbf{bot}(\mathcal{T}_1))$. By Lemma 6.11, the word the a-projection V_a is equal to $V'V''$ where V' is positive, and V'' is negative. Accordingly, we can subdivide $\mathcal{T}_i$ ($i = 1, 2$) into two subbands $\mathcal{T}'_i$ and $\mathcal{T}''_i$ such that

$$\phi(\mathbf{top}(\mathcal{T}'_1))_a = \phi(\mathbf{bot}(\mathcal{T}'_2))_a = V', \phi(\mathbf{top}(\mathcal{T}''_1))_a = \phi(\mathbf{bot}(\mathcal{T}''_2))_a = V'',$$

and we can subdivide Γ into two subdiagrams Γ' and Γ'' where Γ' is formed by the maximal a-bands of Γ starting on $\mathbf{top}(\mathcal{T}'_1)$. Notice that by Lemma 4.5, the length of of each of the a-bands in Γ' is recursively bounded in terms of $|q|$, and the number of these a-bands is recursively bounded as well. Hence the number of cells in Γ' is recursively bounded, and we can assume that $\Gamma = \Gamma''$ (and that the same is true for all other subdiagrams of Δ_1 situated between two consecutive θ-bands).

9. As in the proof of Lemma 9.6, we can deduce that the x-edges in Δ_1 have the same labels and the word V' is a power of some letter $a = a_i(\overleftarrow{L}_j)$. Notice that the letter a does not depend on the choice of Γ. Thus for every maximal θ-band $\mathcal{T}$ of Δ_1, $\phi(\mathbf{bot}(\mathcal{T}))_a = a^k$ or some negative k, $|k| > f(|p|+|q|)$ depending on $\mathcal{T}$. By Lemma 9.3, there exists an a-annulus $\mathcal{C}$ in Δ_1 such that the number of cells between $\mathcal{C}$ and q is recursively bounded. Without loss of generality, we can assume that q is the inner contour of $\mathcal{C}$. Since $k \gg 1$, there exist one more a-annulus $\mathcal{C}_1$ such that the inner contour of $\mathcal{C}_1$ is the outer contour of $\mathcal{C}$.

10. Now suppose that the condition (b) of the lemma holds, that is the base of Δ is $L_j L_j^{-1}$. Then we can repeat the above argument for Δ_2 and conclude that all x-edges of each side of the second L_j-annulus in Δ have the same labels. This allows us to apply Lemma 9.5, eliminate the L_j-annuli from Δ, and then use Lemma 9.2.

11. Finally suppose that the base of Δ is L_j as in case (c) of the lemma.

Notice that the label of the inner boundary of $\mathcal{B}_1$ has the form

$$\theta(x^{-1}a)x^{4k_1}(x^{-1}a)^{-1}\theta^{-1}x^{4k_2}\theta(x^{-1}a)...x^{4k_l}$$
$$= \theta x^{-1}ax^{4k_1}a^{-1}x\theta^{-1}x^{4k_2}\theta x^{-1}a...x^{4k_l}$$

for some integers $k_1, ..., k_l$ where $a \in \mathcal{A}(z_1), x = x(a, \tau)$. The label of the outer boundary of $\mathcal{B}_1$ is freely equal to

$$\theta'(x')^{-1}a'(x')^{k_1}(a')^{-1}x'\theta'^{-1}(x')^{k_2}\theta'(x')^{-1}a'...(x')^{k_l}$$

where θ' is the "brother" of θ, a' is a "brother" of a, $x' = x(a', \tau)$. Notice that for every choice of parameters $k_1, ..., k_l$ there exists an L_j-annulus with boundary labels as above.

Modulo relations (2.8) any word of the form $ax(a,\tau)^t a^{-1}$ is equivalent to $x(a,\tau)^{4t}$. Therefore for every integers $s_1, ..., s_l$ divisible by 16 there exists a roll with base L_j^{-1} and without a-annuli with outer boundary label

$$\theta x^{s_1}\theta^{-1}x^{s_2}\theta x^{s_3}...x^{s_l}$$

and inner boundary label

$$\theta'(x')^{s_1/4}\theta'(x')^{s_2/4}\theta'x^{s_3/4}...x^{s_l/4}.$$

Let us denote this roll by $\Psi(s_1, ..., s_l)$. The "inverse" roll, that is the roll obtained from $\Psi(s_1, ..., s_l)$ by switching the inner and outer contours, will be denoted by $\Psi^{-1}(s_1, ..., s_l)$. The base of the inverse role is L_j.

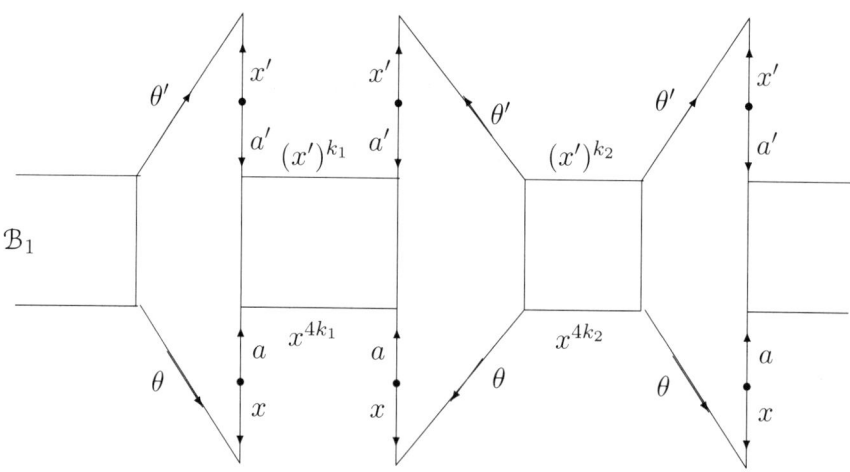

Fig. 27.

Notice that for some integers $s_1, ..., s_l$ divisible by 16 the label of the inner contour of $\Psi^{-1}(s_1, ..., s_l)$ coincides with the label of the outer contour of $\mathcal{C}_1$. Therefore we can do the following surgery: cut the roll Δ along the outer contour of $\mathcal{C}_1$, insert in the hole a copy of $\Psi(s_1, ..., s_l)$ and $\Psi^{-1}(s_1, ..., s_l)$ (the resulting non-reduced "roll" will have the base $L_j L_j^{-1} L_j$). Now reduce the smallest subroll Δ' of the resulting diagram with the base $L_j^{-1} L_j$.

There are two possibilities. Either after the reduction, the base becomes L_j and only the L_j-annulus of $\Psi^{-1}(s_1, ..., s_l)$ will remain, or the base of the new (still non-reduced diagram will be $L_j L_j^{-1} L_j$ but the smallest subroll with base $L_j^{-1} L_j$ will be reduced. Applying Lemma 9.8 to that subroll, we can get back to the first possibility.

Thus we can construct another roll Δ'' with base L_j with the same boundary labels as Δ, such that there are at most two a-annuli between the L_j-annulus of Δ'' and the inner contour of Δ''. By Lemma 9.3 this means that the number of cells between the L_j-annulus of Δ'' and the inner contour is recursively bounded. This implies that the length of the L_j-annulus is recursively bounded in terms of $|q|$, which completes the proof. $\square$

9.3 Rolls with base $L_{j-1}^{-1} K_j L_j$

A cyclic freely trivial word will be called a Dyck word. If W is a Dyck word in the alphabet $\{z_1^{\pm 1}, z_2^{\pm 1}, ...\}$ then for every scheme of cancellation of W, we can associate a pairing of letters of W (i.e. a selection of some pairs of letters in W, each letter occurring in one pair), and a placement of parentheses in W which show in which order we should cancel the letters. This pairing will be called *cancellation pairing* in W. For example, a pairing in the cyclic word $b^{-1}abb^{-1}a^{-1}b$ is represented as $b^{-1})(a(bb^{-1})a^{-1})(b$. Each open parenthesis corresponds to unique closed parenthesis. A pair of open and closed parentheses corresponding to each other will be called *connected*. It is be convenient to draw the word on the boundary of a disc on the plane (we always read cyclic words in the clockwise direction): One can connect the letters of each pair drawn on the disk boundary, by arcs situated on the plane, say, outside the disk, without intersections of these arcs.

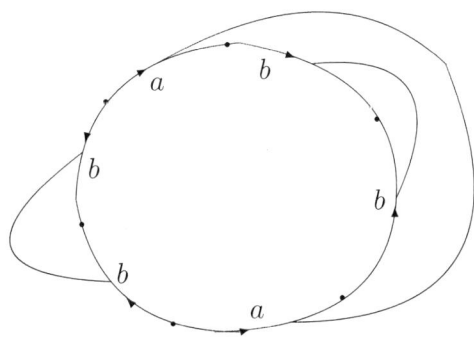

Fig. 28.

Then it is clear when a connected pair of parentheses or letters is *inside* another pair of parentheses (letters) or, vice versa, is *outside*, i.e. *contains* it. From the definition of a cancellation pairing, it is clear that no two open (resp. closed) parentheses can stay next to each other because no letter can cancel twice during the process of cancellation. Thus there is a letter next in the clockwise (counterclockwise) direction of every open (closed) parenthesis. If two parentheses are connected then we call the corresponding letters *connected*.

Let us call a pair of connected letters $(z_i^{\pm 1}, z_i^{\mp 1})$ *normal*, if every pair of connected letters containing this pair consists of different occurrences of the same letter $z_i^{\pm 1}$.

For every cancellation pairing in a cyclic Dyck word W, a pair of connected letters of the form (z_i^{-1}, z_i) will be called *minus pair*, a pair of the form (z_i, z_i^{-1}) will be called a *plus pair*. We call a Dyck word W a *minus word* if for some pairing (called *minus pairing*) every pair of connected letters is a minus pair. Similarly one can define a plus pairing and a plus word.

For example the cyclic word $abb^{-1}a^{-1}abb^{-1}a^{-1}$ is both a plus word:

$$(a(bb^{-1})a^{-1})(a(bb^{-1})a^{-1})$$

and a minus word:

$$a)b)(b^{-1}(a^{-1}a)b)(b^{-1}(a^{-1},$$

but it does not have a pairing where every connected pair of letters is normal.

Lemma 9.10. *If a minus word W contains a subword of the form $z_i^{-1}z_j$, then $i = j$, and the two letters of this subword are connected in any minus-pairing of W.*

Proof. If these two letters are not connected in a given minus pairing then they would be separated by a parenthesis. If is clear that in any minus pairing a negative (positive) letter stays next in the clockwise (resp. counterclockwise) direction of any open (resp. closed) parenthesis. Therefore the parenthesis separating $z_i^{-1}z_j$ must be both open and closed, a contradiction. □

Let A and B be alphabets and let $\phi : A \to B \cup \{1\}$ be a map. Let W be any cyclic group word in the alphabet A. Then any pairing $\mathcal{C}$ in $\phi(W)$ induces a pairing of some letters in W: we simply pull the arrangement of parentheses in $\phi(W)$ to W. Notice that some letters may not be paired in W, and the paired letters may be not mutual inverse, and, in general, the induced pairing is not a cancellation pairing.

Let h be a word over $\mathcal{S} \cup \bar{\mathcal{S}}$. The projection h_0 of h on $\mathcal{S}(2) \cup \mathcal{S}(4)$ will be called the $(2,4)$-*projection* of h. If we further identify in h_0 rules corresponding to the same letter a_i, $i = 1,...,\bar{m}$ (such rules will be called *a-similar*), then we get a word h_1. If h_1 is a Dyck word then any cancellation pairing in h_1 induces a pairing of letters in h. That pairing will be called a $(2,4)$-*pairing* in h. The $(2,4)$-pairing in h will be called z-*good* for some $z \in \tilde{\mathcal{K}}$ if no rule locking zz_+-sectors occurs inside a pair of connected parentheses. A pairing is called z-*best* if no rule locking zz_+-sectors occurs in h. Further, consider any word $W = X_1\theta_1 X_2\theta_2...X_s$ where $\theta_i \in \Theta(z)$, $z \in \{\overleftarrow{L}_j, L_j\}$ for some j, the words $X_1,...,X_s$ do not contain θ-letters. Let h be a history of this word that is the projection of $\theta_1...\theta_{s-1}$ onto $\mathcal{S} \cup \bar{\mathcal{S}}$. Then any $(2,4)$-pairing in h induces a pairing in W. This pairing in W will be called z-*good* (z-*best*) if it is induced by a z-good (z-best) pairing in h.

Lemma 9.11. *Let $z \in \{\overleftarrow{L}_j, L_j\}$ and*

$$w \equiv X_0\theta_1 X_1 \ldots \theta_s X_s \tag{9.1}$$

for some s, where $\theta_i^{\pm 1}$ are letters of the form $\theta(\tau, z)$ for some z, where τ does not lock zz_+-sectors, X_j are words in $\mathcal{X}(z)$. Let $a = a_i(z)$ for some $i \in \{1,...,\bar{m}\}$. Then:

(1) Modulo relations (2.7) and (2.8), we have $awa^{-1} = w'$ where

$$w' \equiv X_0'\theta_1 X_1' \ldots \theta_s X_s',$$

$X_0',\ldots,X_s'$ are words in $\mathcal{X}(z)$. The number of applications of relations (2.7) in the derivation of this equality does not exceed s. If in addition $\theta_t = \theta_{t+1}^{-1} = \theta(\tau, z)^{-1}$ for some index t, and X_t is a power of a letter $x = x(a, \tau)$ then X_t' is again a power of x.

(2) If θ_1 is a negative letter and θ_s is a positive letter then, modulo relations (2.7) and (2.8) the word w is equal to the word

$$w_1 = Y_0(a^{-1}x\theta_1)Y_1\theta_2 Y_2 \ldots \theta_{s-1} Y_{s-1}(\theta_s(x')^{-1}a)Y_s,$$

where $Y_0,\ldots,Y_s$ are words in $\mathcal{X}(z)$, $x = x(a, \tau')$ where $\theta_1^{-1} = \theta(\tau', z)$ and $x' = x(a, \tau'')$ where $\theta_s = \theta(\tau'', z)$. The number of applications of relations (2.7) in the derivation of this equality is at most s. If $\theta_t \equiv \theta_{t+1}^{-1} \equiv \theta_l(\tau, z)^{-1}$, and X_t is a power of $x = x(a, \tau)$, then Y_t is again a power of x.

Proof. (1) By (2.8) $ax = x^4 a$ for $x \in \mathcal{X}(z)$. Furthermore

$$a\theta(\tau, z)^{\pm 1}a^{-1} = x^{\mp 1}\theta(\tau, z)^{\pm 1}x^{\mp 1}$$

by relations (2.7). These relations allow us to move the letter a in awa^{-1} to the right until it cancels with a^{-1}. The additional condition for X_t' can be obtained just as immediately.

(2) First make two insertions which do not change the value of w in the free group:

$$w = X_0 a^{-1}(a\theta_1 X_1 \ldots \theta_s a^{-1})aX_s.$$

Now apply part (1) of the lemma to the word v in parentheses, utilizing the fact that $a\theta(\tau,z))^{-1} = x\theta(\tau,z)^{-1}xa$ by (2.7). As a result, we obtain a word in the desired form. $\square$

Lemma 9.12. *Let Δ be a roll with base K_jL_j or $K_j^{-1}L_{j-1}$ whose history is a word over $\mathcal{S}$. Suppose that the inner contour q of Δ coincides with the inner side q of the K_j-annulus $\mathcal{B}'$ and the outer contour of Δ coincides with the outer contour p of the L_j-annulus $\mathcal{B}$. Then:*

(1) Each maximal a-band in Δ starting on $\mathcal{B}$ ends on $\mathcal{B}$.

(2) If U is the label of the inner contour p_0 of $\mathcal{B}$ then U_a is a Dyck word and the maximal a-bands in Δ determine a cancellation pairing of U_a.

(3) If an a-band $\mathcal{C}$ determines a minus pair of letters (a^{-1}, a) in U_a then (a^{-1}, a) are consecutive letters in U_a, and a subband of $\mathcal{C}$ connects two consecutive θ-cells π_1, π_2 in $\mathcal{B}$. The corresponding subword in the history of the roll is $\tau^{-1}\tau$ where $\tau \in \mathcal{S}^+(2) \cup \mathcal{S}^+(4)$.

(4) Let Γ be a small subtrapezium in Δ with history $\tau^{-1}\tau$ from part (3). Let $\mathcal{T}_1$ and $\mathcal{T}_2$ be two θ-bands containing the cells π_1, π_2. Then the maximal subword in $\mathcal{X}$ written on p_0 (see part (2)) between $\mathcal{T}_1$ and $\mathcal{T}_2$ is a power of a letter $x(a,\tau)$.

Proof. We consider only the case when the base is K_jL_j (the other case is similar).

(1) The statement is obvious since the sides of the K_j-band do not have a-edges.

(2) If an a-band starts on an edge e of the path p_0, $\phi(e) = a^{\pm 1}$ then it ends on an edge of p_0 with label $a^{\mp 1}$. Since the maximal a-bands do not intersect, they determine a pairing in U_a.

(3) Let Δ_0 be the van Kampen subdiagram of Δ bounded by $\mathcal{C}$ and a part of p_0 (and including $\mathcal{C}$). Suppose that Δ_0 contains another maximal a-band $\mathcal{C}'$ corresponding a plus pair. Then we should be able to find a small subtrapezium Γ, whose θ-bands $\mathcal{T}_1, \mathcal{T}_2$ are connected by a subband of $\mathcal{C}$ and by a subband of $\mathcal{C}'$. Then consider the label V of a part of a side of $\mathcal{T}_1$ from $\mathcal{B}'$ to $\mathcal{B}$. When we read the word V from left to right (from $\mathcal{B}'$ to $\mathcal{B}$) then we first read a negative a-letter (since $\mathcal{C}$ corresponds to a minus pair) and then a positive a-letter since $\mathcal{C}'$ corresponds to a plus pair. But this contradicts Lemma 6.11. Hence the a-band $\mathcal{C}'$ does not exist.

It is clear from the form of relations (2.7) that the a-band $\mathcal{C}$ connects two θ-cells π_1 and π_2 corresponding to rules τ_1^{-1} and τ_2 where the rules $\tau_1, \tau_2 \in \mathcal{S}^+(2) \cup \mathcal{S}^+(4)$ are similar. If in the corresponding subword $h = \tau_1^{-1}...\tau_2$ of the history of Δ, there exists a reduced subword of length 2 then there exists a small subtrapezium Γ with reduced history, such that $\mathcal{C}$ crosses Γ. Therefore the word V_0 defined for Γ like the word V in the previous paragraph, contains a negative a-letter. But this contradicts Lemma 6.12.

THE CONJUGACY PROBLEM AND HIGMAN EMBEDDINGS

Therefore $h \equiv \tau^{-1}\tau\tau^{-1}\tau\ldots$. As we have proved before, all pairs of edges on p_0 defined by a-bands in Δ_0 are minus pairs. Obviously it is possible only if $h \equiv \tau^{-1}\tau$.

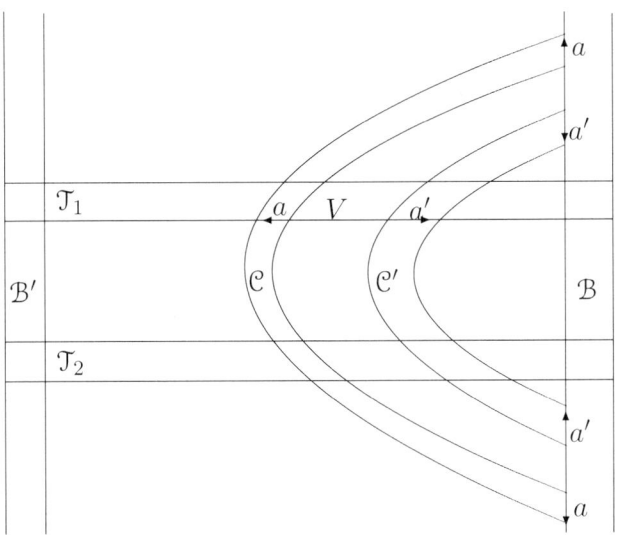

Fig. 29.

(4) By part (3), a subband $\mathcal{C}_0$ of $\mathcal{C}$ connects two a-edges which are the closest a-edges to the L_j-edges of sides of $\mathcal{T}_1$ and $\mathcal{T}_2$. Since $\mathcal{C}$ defines a minus pair in U_a, the word W over $\mathcal{X}$ written on the side of $\mathcal{C}_0$ which is closer to $\mathcal{B}$ is a product of fourth powers (see relations (2.8)). The word W' written on p_0 between $\mathcal{T}_1$ and $\mathcal{T}_2$ is also a product of fourth powers. Then by Lemma 6.1, we have the equality $W' = xWx^{-1}$ in the free group. By Lemma 6.10 (part (4)), it follows that both W and W' are powers of x. □

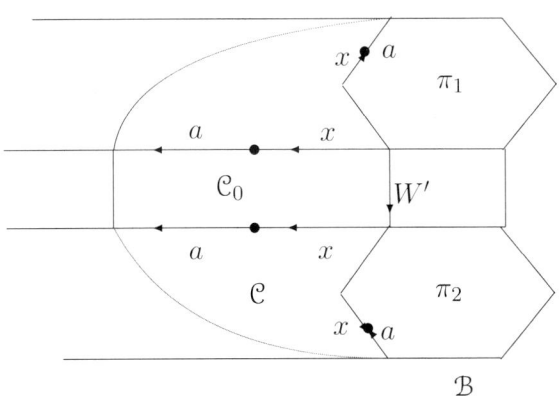

Fig. 30.

Lemma 9.13. *Let Δ be a roll with base L_j, bounded by the inner contour q of an L_j-annulus, and the outer contour p of an a-annulus. Let $\mathfrak{T}$ and $\mathfrak{T}'$ be two consecutive (clockwise) θ-bands connecting q and p. Let Γ be the subdiagram containing $\mathfrak{T}$ and $\mathfrak{T}'$ and all cells between them. Let V be the label of $\mathbf{bot}(\mathfrak{T})$. By Lemma 6.11, $V \equiv V_1 V_2$ where $(V_1)_a$ is positive and $(V_2)_a$ is negative. We claim that either $|V_a| \leq 2|p|$ or $|(V_1)_a| > |p|$.*

Proof. By contradiction, assume that $|(V_2)_a| > |p|$. Let $\mathcal{C}_1, ..., \mathcal{C}_{|p|+1}$ be a series of consecutive a-bands connecting $\mathfrak{T}$ and $\mathfrak{T}'$ enumerated from q to p, so $\mathcal{C}_{|p|+1}$ is a subband of $\mathcal{C}$. Then the length of the word U_1, over $\mathcal{X}$ written on the bottom side of $\mathcal{C}_1$ and the length of the word U_1' written on the top of $\mathcal{C}_1$ are related by the equality $|U_1'| = 4|U_1|$ (see relations (2.8)) because the word $(V_2)_a$ is negative.

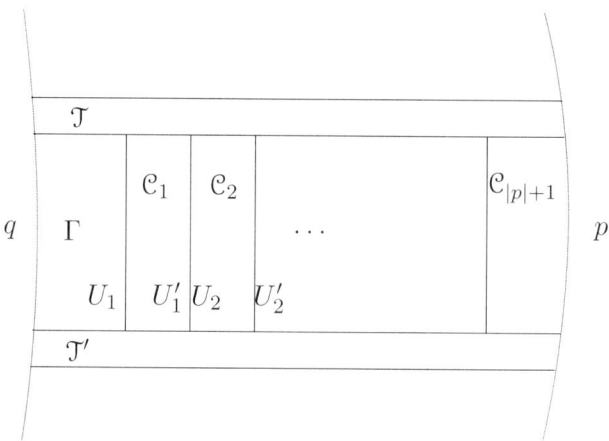

Fig. 31.

Let us introduce similar notation U_2, U_2' for words over $\mathcal{X}$ on the bottom and top sides $\mathcal{C}_2$. Then as in Lemma 9.12 (4), we have $U_2 = x^{\pm 1} U_1' y^{\pm 1}$ for some x-letters x and y. Notice that either U_1 is not empty or $x^{\pm 1} \neq y^{\mp 1}$ because otherwise the (θ, a)-cells connected by $\mathcal{C}_2$ would form a reducible pair of cells. Therefore $|U_2| > |U_1|$. Similarly $|U_2| < |U_3| < ... < |U_{|p|+1}|$. But then $|U_{|p|+1}'| \geq 4|U_{|p|+1}| \geq 4(|p|)$, a contradiction since the definition of $A_{|p|+1}$ implies that $|U_{|p|+1}'| \leq |p|$. $\square$

Lemma 9.14. *In the notation of Lemma 9.13, let the label X of the subpath $\Gamma \cap q$ of q be a power of a letter $x \in \mathcal{X}$, and $\mathcal{C}$ be an arbitrary $a_i(L_j)$-band between $\mathfrak{T}$ and $\mathfrak{T}'$ starting on the "positive" subpath of a side of $\mathfrak{T}$ with label V_1. Then the words in $\mathcal{X}$ on $\partial(\mathcal{C})$ are powers of $x' = x(a_i(L_j), \tau)$, and all a-bands $\mathcal{C}_1, ..., \mathcal{C}_r = \mathcal{C}$ between the L_j-annulus $\mathcal{B}$ and $\mathcal{C}$, connecting $\mathfrak{T}$ and $\mathfrak{T}'$ correspond to the same a-letter $a_i(L_j)$.*

Proof. Let X' be a word in $\mathcal{X}$ written from $\mathcal{T}$ to $\mathcal{T}'$ on the top of the subband $\bar{\mathcal{B}}$ of the L_j-annulus $\mathcal{B}$ whose bottom side is labelled by X. Then relations (2.9) imply that X' is a power of a letter from $\mathcal{X}$. let Y_i and Y_i' are words in $\mathcal{X}$ written respectively on the bottom side (the side that is closer to $\mathcal{B}$) and the top side of the a-band $\mathcal{C}_i$. Then $x_1^{\pm 1} Y_1 x_2^{\pm 1} = X'$ for some x-letters x_1, x_2 since the word $(V_1)_a$ is positive (see Lemma 6.1). Besides Y_1 is a product of fourth powers according to the relations (2.8). This immediately implies that both Y_1 and Y_2 are powers of a letter x' (the assumption that Y_1 is empty and $x_1^{\pm 1}$ and $x_2^{\pm 1}$ are mutually inverse implies that two θ-cells in $\mathcal{T}$, $\mathcal{T}'$ that are connected by $\mathcal{C}_1$ cancel). The letters x_1 and x_2 also must coincide with $(x')^{\pm 1}$. Therefore Y_1 is an a-band where a is determined by the letter x' (see relations (2.7)). Repeating this argument leads to the same conclusion for x-subwords written on the sides of a-bands $\mathcal{C}_2, ..., \mathcal{C}_r = \mathcal{C}$, so all these a-bands correspond to the same letter from $\mathcal{A}$ determined by the x-letter x'. □

Lemma 9.15. *Let a reduced annular diagram Δ consists of one a-annulus with inner side q and outer side p, and the edges with positive labels $a = a_i(L_j)$ cut the diagram from q to p (that is the edges with positive a-labels are oriented outward). Let the word $w \equiv \phi(q)$ have the form (9.1) from Lemma 9.11 for $z = L_j$ and $s > 0$. Suppose that the cyclic word w admits two $(2,4)$-pairings, a L_j-best pairing I and a $\overleftarrow{L_j}$-good pairing II, such that I is a minus pairing, and in II, all minus pairs correspond to occurrences of subwords $\theta^{-1} X \theta$ where X is a power of a letter $x = x(a, \tau)$, $\theta = \theta(\tau, L_j)$. Suppose further that the minus pairs of II are normal pairs for I. Then there exists an annular diagram Δ' with the labels of contours w, w' where w' also has the form (9.1) but for $z = \overleftarrow{L_j}$ and possibly with different x-subwords than w; the diagram Δ' has one L_j-annulus $\mathcal{B}_0$ surrounding the hole of Δ', other cells in Δ' correspond to relations (2.7) and (2.8), and the number of θ-cells in Δ' is less than s^2.*

Proof. By Lemma 9.11 (2), using the pairing I we can replace each pair of connected letters $\ldots \theta^{-1} \ldots \theta' \ldots$ in the cyclic word w by a pair of subwords $\ldots (a^{-1} x \theta^{-1}) \ldots (\theta' x'^{-1} a) \ldots$ (starting with the outermost pairs in I). The number of (θ, a)-relations used in the transition to the new word w_1 is less than $s^2/2$. The form of the word w_1 given in Lemma 9.11 (2) and the form of relations (2.6) corresponding to rules from $\mathcal{S}(2) \cup \mathcal{S}(4)$, and relations (2.7) allows us to glue an L_j-annulus $\mathcal{B}_0$ to the annular diagram of conjugacy of w and w_1 along the path with label w_1 (the top side of $\mathcal{B}_0$).

In the transition from w to w_1, we do not change the θ-letters. Suppose that an occurrence of a subword of the form $\theta^{-1} X \theta$ corresponds to a minus pair in II. By Lemma 9.10 this pair is connected in I as well. This pair is normal in I by the assumption of the lemma. Therefore in the subword $(a^{-1} x \theta^{-1}) Y (\theta x^{-1} a)$ of the word w_1 obtained by Lemma 9.11 (2) from the subword $\theta^{-1}X\theta$ of w, the word Y is a power of the same letter x.

Let w_2 be the word written on the bottom side of our L_j-annulus $\mathcal{B}_0$. The subword of w_1 discussed in the previous paragraph corresponds to the subword of the form $(a_-^{-1}x_-\theta_-^{-1})Y_-^4(\theta_-x_-^{-1}a_-)$ in w_2 (see relations (2.9) and (2.6)) where the minus sign in the subscript means that we replace L_j by $\overleftarrow{L}_j$ in $a_i(L_j)$, $x(a_i(L_j),\tau)$ and $\theta(\tau,L_j)$. Since we have a fourth power in the middle of this word, we can transform this occurrence into $\theta_-^{-1}Y_-\theta_-$ without a-letters applying relations (2.8) and two relations (2.7).

In the resulting word w_3, a-letters occur only in subwords of the form $(\theta x^{-1}a)^{\pm 1}$, where the letter θ occurs in a plus pair of the pairing II. These a-letters can be removed using relations (2.7) and (2.8) by Lemma 9.11 (1). As a result, we get a word $w_4 = w'$ and the diagram Δ' consists of a L_j-annulus $\mathcal{B}_0$ and other cells described above. This diagram satisfies all the conditions of the lemma. $\square$

Lemma 9.16. *Let Δ be a roll with base K_jL_j or $K_j^{-1}L_{j-1}$ (j is odd), whose history is not empty and does not contain rules from $\overline{\mathcal{S}}$. Suppose the outer contour p of the roll is the outer contour of some $a_i(L_j)$-annulus. Then either between the L_j-annulus $\mathcal{B}$ and p, there exists an a-annulus $\mathcal{C}$ satisfying the conditions of Lemma 9.15 and the length of every θ-band connecting $\mathcal{B}$ and $\mathcal{C}$ is at least $|p|$ or $\mathcal{B}$ and p are connected by a θ-band of length $\leq 2|p|$.*

Proof. We can consider only the case when the base is K_jL_j. Let Δ_1 be the annular subdiagram in Δ bounded by $\mathcal{B}$ and p. By Lemma 9.13 we can assume that for every two consequent θ-bands $\mathcal{T}$ and $\mathcal{T}'$ connecting $\mathcal{B}$ and p, $|V_1|_a > |p|$ in the notation of Lemma 9.13. If we enumerate all a-annuli between $\mathcal{B}$ and p from $\mathcal{B}$ to p: $\mathcal{C}_1, \mathcal{C}_2, \ldots$, then we can let $\mathcal{C}$ be the first a-annulus which cannot be connected with $\mathcal{B}$ by any θ-band of length $< |p|$.

Since Δ_1 has an a-annulus by the lemma assumption, we have that every maximal a-band $\mathcal{C}'$ in Δ_1 that starts on $\mathcal{B}$, ends on $\mathcal{B}$, and we obtain a pairing in the word U'_a where U' is written on the top side of $\mathcal{B}$.

Since the a-letters in a relation (2.6) corresponding to rules from $\mathcal{S}(2) \cup \mathcal{S}(4)$ determine the rule up to the similarity relation, we have a $(2,4)$-pairing I in the word U'.

Any a-band $\mathcal{C}'$ of Δ_1, starting and ending on $\mathcal{B}$ has a subband connecting a pair of consecutive θ-bands $\mathcal{T}$ and $\mathcal{T}'$. Moreover $\mathcal{C}'$ must intersect the path labelled by V_1 (in the above notation) because between $\mathcal{C}'$ and $\mathcal{T}$ there are less than $|p|$ maximal a-bands and $|V_1|_a \geq |p|$. Since $(V_1)_a$ is a positive word, $\mathcal{C}'$ determines a minus pair of a-letters. Therefore I is a minus-pairing. Notice that this pairing is L_j-best. Indeed, every maximal θ-band in Δ intersects the annulus $\mathcal{C}$. Hence for each θ-letter on the outer contour of $\mathcal{B}$, there exists a (θ, a)-relation of the form (2.7) containing θ. Thus the corresponding rule from $\mathcal{S}$ does not lock L_jP_j-sectors.

THE CONJUGACY PROBLEM AND HIGMAN EMBEDDINGS 117

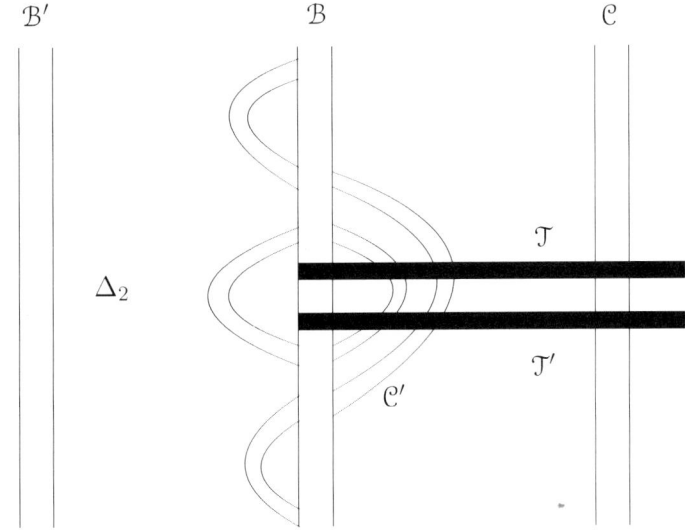

Fig. 32.

Now consider the subroll Δ_2 bounded by $\mathcal{B}$ and the K_j-annulus $\mathcal{B}'$. Lemma 9.12 (2) gives a pairing in the word U_a where U is the label of the bottom of $\mathcal{B}$. This automatically gives a $(2,4)$-pairing in the word U which we shall denote by II. This pairing is K_j-good because an a-band that connects two edges on the inner contour of $\mathcal{B}$ cannot cross a θ-band corresponding to a rule locking $K_j L_j$-sectors (see relations (2.7)). By Lemma 9.12 (3,4) every minus pair of θ-letters of II corresponds to an occurrence of a subword $\theta^{-1} Y \theta$ in U where Y is a power of an x-letter $x = x(a, \tau)$ (a is obviously determined by θ by the relations (2.6)). By Lemma 9.10 the same pair of letters must be connected in the pairing I, and by Lemma 9.14 for Δ_1 the corresponding occurrence in $\phi(p)$ also has the form $\theta^{-1} X \theta$ where X is a power of the letter $x(a', \tau)$ where a' is the "brother" of a. The normality condition of the bottom of $\mathcal{C}$ follows from Lemma 9.14 and the choice of $\mathcal{C}$. □

Lemma 9.17. *Let Δ be a roll with base $L_{j-1}^{-1} K_j L_j$ for some j and history h, $|h| > 0$. Let the outer contour p of Δ be the outer contour of some a-annulus, and h does not contain rules from $\bar{\mathcal{S}}$. Then there exists a roll $\bar{\Delta}$ with the same boundary labels and the same base as Δ where the L_j-annulus can be connected with p by a θ-band of length $\leq 2|p|$.*

Proof. By Lemma 9.16 applied to the subroll with base $K_j L_j$ we can assume that there exists an a-annulus $\mathcal{C}$ between the L_j-annulus $\mathcal{B}$ and p satisfying the conditions of Lemmas 9.16 and 9.15. Hence the number of (θ, a)-cells between $\mathcal{B}$ and $\mathcal{C}$ is at least $|p|s$ which, in turn, at least s^2 where s is the length of the history of the roll.

Consider the auxiliary diagram Δ' corresponding to $\mathcal{C}$ as in the conclusion of Lemma 9.15. Let Δ'' be the diagram obtained from Δ' by first taking the mirror image and then changing the indexes L_j to L_{j-1}, and K_j to K_j^{-1} in all labels (Lemma 3.13 allows us to do that). Lemma 9.15 and the form of relations (2.6) corresponding to K_j imply that there exists a K_j-band $\mathcal{B}'$ whose outer contour coincides with the inner contour of Δ' and whose inner contour coincides with the outer contour of Δ''. The diagram Δ_0 obtained by gluing these three diagrams ((Δ', $\mathcal{B}'$ and Δ'') has outer contour q_0 labelled by w (same as $\mathcal{C}$) and inner contour labelled by the word w'' obtained from w by replacing the index L_j with L_{j-1} in all letters.

Therefore we can do the following transformation with Δ. First cut Δ along q_0 (the inner contour of $\mathcal{C}$), along the outer contour p_1 of the K_j-annulus $\mathcal{B}''$ of Δ, and along the inner contour q_1 of this annulus. We get three annular diagrams Δ_1, Δ_2 and Δ_3 where Δ_1 is bounded by p and q_0, Δ_2 is bounded by q_0 and p_0 and Δ_3 is bounded by q_1 and q. Then change index K_j to L_{j-1}^{-1}, L_j by L_{j-1} in all labels of edges of Δ_2. Let Δ_2' be the resulting diagram. The label of the outer contour of Δ_2' is identical with the label of the inner contour of Δ_0 and the label of the inner contour of Δ_2' coincides with the label of q_1 as follows from the form of K_j-relations (2.6) and (2.9). Let $\bar{\bar{\Delta}}$ be the result of gluing $\Delta_1, \Delta_0, \Delta_2'$, and Δ_3 and reducing the resulting diagram. By Lemma 9.15, the number of (θ, a)-cells between the L_j-annulus of $\bar{\bar{\Delta}}$ and p is smaller than the similar number for Δ. So our transformation moved the L_j-annulus "closer" to p.

The base of $\bar{\bar{\Delta}}$ before reducing was $L_{j-1}^{-1}L_{j-1}L_{j-1}^{-1}K_jL_j$ (one L_{j-1}-band comes from Δ_1, and two from Δ_0). After reducing two of the L_{j-1}-bands can disappear.

If this does not happen then the history of $\bar{\bar{\Delta}}$ is a word over $\mathcal{S}(2) \cup \mathcal{S}(4)$ (since other rules lock $L_{j-1}^{-1}K_j$-sectors or $P_{j-1}^{-1}L_{j-1}^{-1}$-sectors). Hence we can apply Lemma 9.8 and remove two of the L_{j-1}-annuli from $\bar{\bar{\Delta}}$.

Thus after the transformations described above we get a roll with the same base, same labels of the contours as Δ but with the L_j-annulus closer to p than in Δ. Moving the L_j-annulus further toward p, we obtain the desired roll $\bar{\Delta}$. $\square$

9.4 Arbitrary rolls

Lemma 9.18. *Let Δ be a roll with contours p and q. Then there exists a roll Δ' with the same boundary labels as Δ, which contains a path t connecting the two contours of Δ', whose length is recursively bounded in terms of $|p| + |q|$.*

Proof. By Lemma 5.5, we can assume that Δ satisfies (R1) and (R2).

If the history of Δ is empty then Δ contains only $\mathcal{G}$-cells and auxiliary cells corresponding to relations (2.8). Moreover, a $\mathcal{G}$-cell cannot have common edges with the $(\mathcal{A}, \mathcal{X})$-cells. Thus Δ must be a union of $\mathcal{G}$-cells and diagrams over $\mathcal{H}_2$, and different components of this union do not have common edges, i.e, Δ is the diagram over the free product of groups. Hence the statement of the lemma in that case follows from the solvability of the conjugacy problem in $\mathcal{G}$ and the conjugacy problem in $\mathcal{H}_2$ (Lemma 4.4).

Hence we can assume that the history of Δ is not empty. Let $\mathfrak{T}_1, \mathfrak{T}_2, \ldots$ be the maximal θ-bands in Δ connecting q and p.

Denote by $\mathcal{B}_1, \ldots, \mathcal{B}_r$ the basic annuli of the roll Δ counted from q to p. We are going to recursively bound the number r as function of $|p|, |q|$ in Δ or in a roll with the same boundary labels and to use induction on r. The case $r = 0$ has been treated in Lemma 9.2.

Recall that there are no auxiliary x-cells in P- or R-bands. Hence the length of every P- or R-annulus in Δ is at most $\min(|p|, |q|)$. One may assume that there are no such distinct bands with equal boundary labels, because this would allow us to delete an annular subdiagram of Δ and to reduce the number of k-annuli in Δ. Since the number and the lengths of P- and R-annuli are recursively bounded, our task is reduced to rolls having nether P- nor R-annuli.

Assume that there is a K_j-annulus $\mathcal{B}_s$ and a $K_{j'}$-annulus $\mathcal{B}_{s'}$ where $s' > s$. We choose s' so that $s' - s$ is minimal possible. Since there are neither P- nor R- annuli among $\mathcal{B}_s, \mathcal{B}_{s+1}, \ldots, \mathcal{B}_{s'}$, we have $j' = j$ by Lemma 6.2, and the base of the subroll Δ' of Δ bounded by $\mathcal{B}_s$ and $\mathcal{B}_{s'}$ has one of the forms:

$$K_j L_j L_j^{-1} L_j \ldots L_j^{-1} K_j^{-1}, \tag{9.2}$$

or

$$K_j^{-1} L_{j-1} L_{j-1}^{-1} \ldots K_j, \tag{9.3}$$

if j is odd, or

$$K_j K_j^{-1} \text{ or } K_j^{-1} K_j \tag{9.4}$$

if j is even.

We shall assume that j is odd, the other case is similar. By Lemma 6.2 none of the rules in the history of the roll locks $L_j P_j$-sectors if the base of Δ' has the form (9.2), and none of them locks $L_{j-1} P_{j-1}$-sectors if the base has the form (9.3). In particular, if the history of the roll contains rules from $\bar{\mathcal{S}}$, then $j \neq 1$ in the (9.2) case. In addition, if there is a $\bar{\Theta}$-band in Δ', then there are no Θ-bands, because a common admissible word for $\mathcal{S}$ and $\bar{\mathcal{S}}$ have no $a(L_j)$-letters, and so, for any j, it has no subwords with base $L_j L_j^{-1}$.

The subroll bounded by $\mathcal{B}_s$ and $\mathcal{B}_{s'}$ has no $\mathcal{G}$-cells by Lemma 6.1 and by Lemma 6.5 applied to the small subtrapezia of Δ'. By Lemma 9.1, all a-edges in the subroll between $\mathcal{B}_t$ and $\mathcal{B}_{t+1}$, $s \leq t \leq s' - 1$ have labels from $\mathcal{A}(z) \cup \bar{\mathcal{A}}(z)$ where $z = z(t)$ depends on t only (z is either K_j or L_j if the base of Δ' has the form (9.2) and z is either K_j^{-1} or L_{j-1} if the base has the form (9.3).

We are going to show how to eliminate the two K_j-annuli from Δ without changing the boundary labels of Δ. Suppose that the base has the form (9.2). Recall that by Lemma 6.2, there are no $\bar\Theta$-cells in Δ' if $j = 1$. Therefore every auxiliary relation involving letters from $\mathcal{A}(K_j) \cup \bar{\mathcal{A}}(K_j) \cup \Theta(K_j) \cup \bar\Theta(K_j) \cup \mathcal{X}(K_j)$ and involved in Δ', has a copy involving letters from $\mathcal{A}(L_{j-1}^{-1}) \cup \bar{\mathcal{A}}(L_{j-1}^{-1}) \cup \Theta(L_{j-1}^{-1}) \cup \bar\Theta(L_{j-1}^{-1}) \cup \mathcal{X}(L_{j-1}^{-1})$, every L_j-cell has a copy L_{j-1}-cell, and every auxiliary relation involving letters from $\mathcal{A}(L_j) \cup \bar{\mathcal{A}}(L_j) \cup \Theta(L_j) \cup \bar\Theta(L_j) \cup \mathcal{X}(L_j)$ (and involved in Δ') has a copy involving letters from $\mathcal{A}(L_{j-1}) \cup \bar{\mathcal{A}}(L_{j-1}) \cup \Theta(L_{j-1}) \cup \bar\Theta(L_{j-1}) \cup \mathcal{X}(L_{j-1})$. Notice that the labels of the inner and outer side of a K_j-band become identical after a substitution of each letter from $\Theta(K_j) \cup \bar\Theta(K_j) \cup \mathcal{X}(K_j)$ by its "brother" in $\Theta(L_{j-1}^{-1}) \cup \bar\Theta(L_{j-1}^{-1}) \cup \mathcal{X}(L_{j-1}^{-1})$ (see Lemma 3.13).

Let us replace every label in the subroll of Δ, bounded by the outer side of $\mathcal{B}_s$ and the inner side of $\mathcal{B}_{s'}$, by its copy in the sense of the previous paragraph. Then we obtain a roll whose boundary labels coincide with the boundary labels of the subroll Δ'. Hence we can replace Δ' in Δ by a subroll with fewer K_j-annuli, as desired. Let us call this operation *the operation of removing K_j-annuli*.

In the case when the base has the form (9.3) the procedure of removing K_j-bands is similar. The only exception is the case when $j = 1$. In that case $(L_{j-1}, \bar\Theta)$-cells do not have L_1-copies. But then all θ-cells of Δ' are $\bar\Theta$-cells, as we noticed earlier. So we need to apply the homomorphism that kills $\bar{(a)}K_j^{-1}$- and $\bar{a}(L_{j-1})$-letters. This homomorphism sends relations to relations, so the construction carries out without other changes.

Hence Δ contains at most one K_j-annulus. Suppose Δ does not have K_j-annuli and the base of Δ has at least 3 letters. Then the history of Δ is a words in $\mathcal{S}(2) \cup \mathcal{S}(4) \cup \bar{\mathcal{S}}$ since other rules lock either $L_j P_j$- or $L_j^{-1}\bar{L}_j^{-1}$-sectors. In this case the base can be shorten by Lemma 9.8. If there are K_j-annuli in Δ and its base has at most two letters, then we can complete the proof by applying Lemma 9.9. So let us assume that Δ contains exactly one K_j-annulus $\mathcal{B}_s$. Thus (up to an inside-out transformation of the roll) the base of Δ has the form $...L_{j-1}L_{j-1}^{-1}K_jL_jL_j^{-1}...$ if j is odd or K_j if j is even (the L_j's may be absent even if j is odd).

If the base of Δ contains a subwords $L_s L_s^{-1}$ and $L_s^{-1} L_s$ then none of the rules in the history of Δ locks $L_s^{-1}z$- or L_s-sectors. Hence the history of Δ is a word in $\mathcal{S}(2) \cup \mathcal{S}(4) \cup \bar{\mathcal{S}}$. By Lemma 9.8, every subroll with base $L_s^{-1}L_s$ can be replaced by a subroll with an empty base. Hence we can assume that the base of Δ does not contain subwords $L_s^{-1}L_s$. Therefore the base of Δ has length at most 5.

Thus k-annuli of Δ divide Δ into at most six subrolls with empty bases. Let us number these subrolls $\Delta_1, \Delta_2, ..., \Delta_{r+1}$, $r \leq 5$, counting from q to p. As in Lemma 9.9, let us fix a "big enough" recursive function $f(x)$, it will be clear later, how big this function should be.

If Δ_1 or Δ_{r+1} has a path of length $\leq f(|p|+|q|)$ connecting its boundary components then using Lemma 5.3, we can show that the number of cells in Δ_1 or Δ_{r+1} is recursively bounded in terms of $|p|+|q|$, and we can remove these subrolls and $\mathcal{B}_1$ or $\mathcal{B}_{t-1}$ from Δ reducing the number of k-annuli in Δ. Thus we can assume that Δ_1 and Δ_{r+1} have no short cuts.

Suppose that the base of Δ starts with K_j. By Lemma 9.3, the roll Δ_1 contains an a-annulus $\mathcal{C}$. It is easy to see that one can build a K_j-annulus $\mathcal{B}'$ whose outer side label coincides with a inner side label of $\mathcal{C}$. Now cut the roll Δ along that side of $\mathcal{C}$, and insert in the hole the K_j-band $\mathcal{B}'$ and its inverse (so that the two K_j-bands cancel each other). As a result we get a non-reduced annular diagram with three K_j-annuli. Using the operation of removing K_j-annuli, we can now remove the bands $\mathcal{B}_1$, $\mathcal{B}'$. After reducing the new annular diagram, we get a roll with the same boundary labels and the same base as Δ but with the K_j-band closer to q than in Δ. Let us call this operation *the operation of moving a K_j-band*. We can use this operation to move the K_j-band towards q until there are no more a-annuli between it and q. By Lemma 9.3, if that was the case then the number of cells between this K_j-band and q is recursively bounded, so we would be able to remove the subroll bounded by the K_j-band and q from Δ, and then apply Lemma 9.9. Similarly we can treat the case when the base of the roll ends with K_j.

Suppose that the base of Δ starts with $L_{j-1}L_{j-1}^{-1}K_j$. Then again by Lemma 9.3 Δ_1 contains an a-annulus $\mathcal{C}$ whose inner side label is the outer side label of some K_j-band. Hence we can do the operation of removing a K_j-band. As a result of this operation we obtain a roll whose base starts with K_j (and contains only one occurrence of $K_j^{\pm 1}$), so we reduce the problem to the previous case. Similarly we can argue when the base of Δ ends with $L_j L_j^{-1}$.

Hence we can assume that the base of Δ has the form $L_{j-1}^{-1}K_j L_j$. As before, we can assume that the number of a-annuli in Δ between p and the L_j-annulus, and between q and the L_{j-1}-annulus is big enough (in particular, > 1). This implies that the history of the roll cannot contain rules that lock $L_j P_j$-sectors. The history of Δ cannot contain both rules from $\mathcal{S}$ and rules from $\bar{\mathcal{S}}$ because no a-letters are both of the form $a_i(L_j)$ and $\bar{a}(L_j)$. If the history is a word over $\mathcal{S}$, we can apply Lemma 9.17 and move the L_j-annulus within a recursive distance from p, and then eliminate this annulus from the diagram reducing the base.

Hence we can assume that the history is a word in $\bar{\mathcal{S}}$. Since there exist no $(\bar{a}, x)$-relations, every a-annulus of Δ does not contain x-edges, and the label of the outer side of it consists of θ-edges. Hence the lengths of all a-annuli in Δ are bounded by $|p|+|q|$. Hence if the number of a-annuli in Δ is big enough, there exist two a-annuli with the same labels of outer contours. Then we can reduce the number of a-annuli by removing the one of the two a-annuli and all cells between them. $\square$

10 Arrangement of hubs

Lemma 10.1. *Let a reduced diagram Δ have exactly one hub Π. Assume that a K_j-band $\mathcal{B}_j$ and a K_{j+1}-band $\mathcal{B}_{j+1}$ start on Δ for $j \neq 1$, and a θ-band $\mathcal{T}$ cross these K-bands and does not cross any other K-bands. Assume further that $\partial \Delta = p(j)q(j)t(j)q'(j)$ where $p(j)$ is a subpath of $\partial \Pi$, $q(j)$ and $q'(j)$ are top or bottom paths of $\mathcal{B}_j$ and $\mathcal{B}_{j+1}$, respectively, and $t(j)$ is the top of $\mathcal{T}$.*

Then one can attach a diagram over $\mathcal{H}_1$ to Δ along $q'(j)tq(j)$ and obtain a diagram Δ' whose perimeter is bounded from above by a linear function of the length of $\mathcal{T}$. In addition, $\phi(\partial \Delta') \equiv \Sigma$ if $\phi(t(j))_a$ is empty.

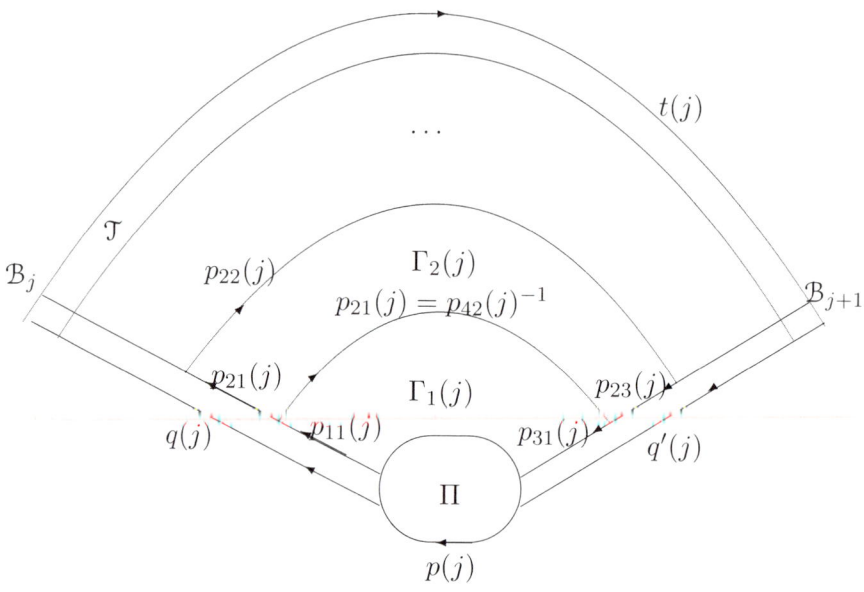

Fig. 33.

Proof. Let $\bar{\Delta}$ be the diagram obtained from Δ after deleting of the hub Π. Obviously, $\bar{\Delta}$ is a trapezium with base $\overleftarrow{L_j}L_jP_jR_j\overrightarrow{R_j}$. If we delete $\mathcal{B}_j$ and $\mathcal{B}_{j+1}$ from $\bar{\Delta}$, we obtain a diagram $\Gamma(j)$ with contour $p_{\Gamma(j)}q_{\Gamma(j)}t_{\Gamma(j)}q'_{\Gamma(j)}$ where $q_{\Gamma(j)}$ ($q'_{\Gamma(j)}$) is a the inner top/bottom of $\mathcal{B}_j$ (of $\mathcal{B}_{j+1}$), $p_{\Gamma(j)}$ is a subpath of $\partial \Pi$ and $t_{\Gamma(j)}$ is a subpath of $t(j)$. Our goal is to construct similar diagrams $\Gamma(j')$ for other indices j' so that the labels of $p_{\Gamma(j')}$, $q_{\Gamma(j')}$ and $q'_{\Gamma(j')}$ have index j' and they are the copies of the labels of $p_{\Gamma(j)}$, $q_{\Gamma(j)}$ and $q'_{\Gamma(j)}$, respectively, and $|t_{\Gamma(j')}| \leq |t_{\Gamma(j)}|$. Then we can attach every such $\Gamma(j')$ (or its mirror copy if $j'-j$ is odd) to Δ and insert copies of $\mathcal{B}_j$ (if $j'-j$ is even) or of $\mathcal{B}_{j+1}$ (if $j'-j$ is odd) between the copies of $\Gamma(j)$. Then we obtain the desired diagram Δ'.

To construct $\Gamma(j')$ we subdivide the trapezium $\bar{\Delta}$ into subtrapezia Δ_1, $\Delta_2, \ldots$ of the first and the second types by making cuts along the boundaries of the maximal $\bar{\Theta}$-bands of $\bar{\Delta}$ so that the subtrapezia of the first and the second types alternate in $\bar{\Delta}$. The same cuts partition $\Gamma(j)$ into subdiagrams $\Gamma_1(j), \Gamma_2(j), \ldots$, when counting from Π to t. The boundary of $\Gamma_s(j)$ is of the form $p_{1s}(j)p_{2s}(j)p_{3s}(j)p_{4s}(j)$ where $p_{2s}(j) = p_{4,s+1}(j)^{-1}$ and the paths $p_{1s}(j)$ (the paths $p_{3s}(j)$) are subpaths of $q_{\Gamma(j)}$ (of $q'_{\Gamma(j)}$).

Notice that $a(\overleftarrow{L_j})$-, $a(L_j)$- or $a(R_j)$-bands of a subtrapezia Δ_s of the second type cannot start on the boundary of Δ_s because the common letters of $\mathcal{A}$ and $\bar{\mathcal{A}}$ belong to $\cup_j \mathcal{A}(P_j)$. Hence such bands contain only $\bar{a}$-edges, and consequently, Δ_s has no (a, x)-cells.

Since the boundary label of every $\Gamma_s(j)$ obviously has index j, we can apply Lemma 3.13 or Lemma 3.14, (depending on the type of trapezia Δ_s) and obtain diagrams $\Gamma_s(j')$ whose boundaries $p_{1s}(j')p_{2s}(j')p_{3s}(j')p_{4s}(j')$ are labelled by words having index j'; the labels of $p_{1s}(j')$ and $p_{3s}(j')$ are copies of the labels of $p_{1s}(j)$ and $p_{3s}(j)$, respectively. The lengths of $p_{2s}(j')$ and $p_{4s}(j')$ are not greater of those for index j, but to construct the desired diagram $\Gamma(j')$ from $\Gamma_s(j')$, $s = 1, 2, \ldots$ (and from some auxiliary cells) we have to compare the labels of the $p_{2s}(j')$ and $p_{4,s+1}(j')$ for $j' = 1$. (They are equal for $j' \neq 1$.) We will assume that $s = 1$, and Δ_1 is of the first type, because other comparisons are similar.

The word $\phi(p_{21}(1))$ is the copy of $V \equiv \phi(p_{21}(j)) \equiv \phi(p_{42}(j)^{-1})$ by the definition of the mapping ε_1, and $\phi(p_{42}(1))$ is obtained from V by deleting all a-letters. However, all a-letters of V are contained in the subword U in $\bar{a}_1, \ldots, \bar{a}_m$ written between the P_j- and the R_j-letters of V (see lemma 6.1) because $\mathcal{A}$ and $\bar{\mathcal{A}}$ have no other common letters with index j. We are going to prove that $U = 1$ modulo $\mathcal{G}$-relations. This will allow us to construct an auxiliary intermediate diagram consisting of $\mathcal{G}$-cells and having the boundary label $p_{21}(1)p_{42}(1)$. Thus the desired diagram $\Gamma(1)$ will be built from $\Gamma_1(1), \Gamma_2(1), \ldots$ and the intermediate diagrams consisting of $\mathcal{G}$-cells.

It remains to show that $U = 1 (\mathrm{mod}\ \mathcal{G})$. For this purpose we apply homomorphism δ from Lemma 3.1 to the boundary label W_1 of $\Gamma_1(j)$. Since the all a-letters of W_1 occur in the subword U, we have $U = 1$ in $\bar{\mathcal{G}}$. By Lemma 3.9, $U = 1$ modulo $\mathcal{G}$-relations as desired. (When arguing in this way on a diagram $\Gamma_s(j)$, $s > 1$, we have to induct on s and take into account that, by the inductive hypothesis, δ sends the complement of U in the boundary label W_s of $\Gamma_s(j)$ to 1.)

The last statement of the lemma clearly follows from the construction of Δ'. $\square$

Lemma 10.2. *Assume that a Van Kampen or an annular diagram Δ over the group $\mathcal{H}$ contains two hubs, Π_1 and Π_2, and two bands, a K_j-band $\mathcal{B}_1$ and a K_{j+1}-band $\mathcal{B}_2$ connecting the hubs where $j \neq 1$. Suppose that the diagram bounded by the two k-bands and the boundaries of Π_1, Π_2 does not* *contain any other hubs and does not contain the hole of the diagram (in the annular case). Then the diagram Δ is not minimal: one can decrease the number of hubs by 2 while preserving the boundary label.*

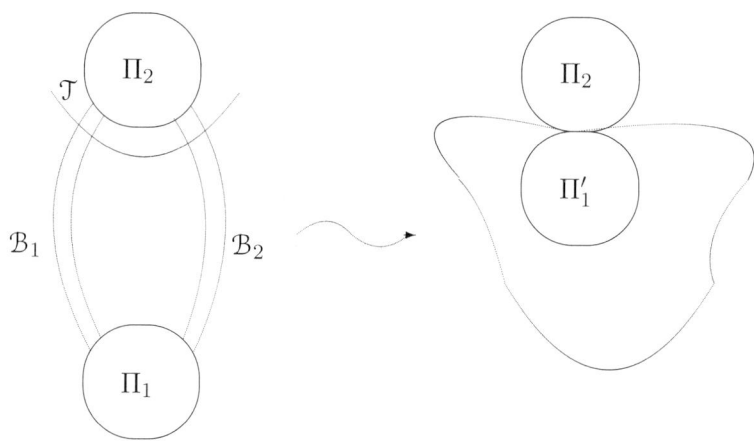

Fig. 34.

Proof. We can assume that there are no cells in Δ distinct from the hubs, the cells of the two k-bands, and the cells entirely lying between the two k-bands. In particular, Δ is simply connected. We can assume also that the subdiagram Δ_0 that consists of the same cells with the exception of the hubs is minimal.

It is clear that Δ_0 is a trapezium with base $\overleftarrow{L_j}L_jP_jR_j\overrightarrow{R_j}$ or it has no θ-cells. In the last case the P_j-edges of Π_1 and Π_2 must coincide. Thus Π_1 and Π_2 cancel as desired.

Otherwise we consider the nearest θ-band $\mathcal{T}$ to Π_2. Its top has no a-edges because otherwise an a-band starting on $\mathcal{T}$ must end on the hub Π_2, a contradiction. Then we apply Lemma 10.1 to the diagram bounded by $\Pi_1, \mathcal{B}_1, \mathcal{B}_2$ and $\mathcal{T}$. It says that we can attach a subdiagram Γ over $\mathcal{H}_1$ and its mirror copy Γ' to Δ (preserving the boundary label of Δ) so that Γ together with Δ_0 and Π_1 form a subdiagram Π_1' whose boundary label is equal to Σ. Therefore we can change Δ by a subdiagram with two hubs which have a common P_j-edge. Then we cancel them as in the first case. $\square$

We extend the notion of *l-graph* used in [Ol2], [SBR], [BORS], [OlSa2], [OlSa1], because here we need such planar graphs not only in a plane but also in an annulus. In the annular case, we have to introduce 2 exterior vertices v_0 and v^0 lying respectively in the outer and in the inner plane components of the complement of the annulus. By definition of l-graph Γ, each of interior vertices $v_1, \ldots, v_n$ is of degree at least l, where $l \geq 6$, $n \geq 1$, and Γ has neither loops, nor 2-gons for which both vertices are interior.

As in [Ol2], [SBR], [BORS] one easily obtains the following corollary of the Euler formula for planar graphs.

Lemma 10.3. *For every l-graph Γ there exists an interior vertex o of degree $d \geq l$ connected with v_0 by d_0 successive edges (such that there are no vertices of Γ between them) and with v^0 by d^0 successive edges where $d_0 + d^0 \geq d - 3$. The number of vertices n does not exceed the sum of degrees of v_0 and v^0 if $l \geq 7$.*

Lemma 10.3 can be applied to the hub graph $\Gamma(\Delta)$ of a minimal diagram Δ which contains hubs. By definition each the interior vertices of $\Gamma(\Delta)$ are chosen inside all hubs, and the exterior vertex (vertices) are chosen in the connected components of the complement of Δ (one vertex per component) on the plane. The medians of all the K_j-bands of Δ ($j \neq 1$) starting on hubs serve as the edges of $\Gamma(\Delta)$.

The next Lemma is similar to Lemma 11.4 in [SBR], Lemma 9.4 [Ol2] and Lemma 4.23 in [BORS]. It immediately follows from Lemma 10.2.

Lemma 10.4. *If a minimal diagram over $\mathcal{H}$ contains hubs, then the graph $\Gamma(\Delta)$ is an $(N-1)$-graph.*

Lemma 10.5. *Let Δ be an annular diagram over $\mathcal{H}$ with contours p and p' or a van Kampen diagram with contour p. Assume Δ contains a hub Π with the following properties.*

1. There exists a K_j-band $\mathcal{B}$ and a K_{j+1}-band $\mathcal{B}'$ connecting $\partial(\Pi)$ with p, $j \neq 1$.

2. The van Kampen subdiagram Γ, which is bounded by p, Π and $\mathcal{B}, \mathcal{B}'$ and includes $\mathcal{B}, \mathcal{B}'$ is reduced and contains no hubs.

Then the length of θ-band $\mathcal{T}$ that is closest to p and crosses both $\mathcal{B}$ and $\mathcal{B}'$ (if any exists) is $O(|p|^2)$, and there is a path t of length $O(|p|)$ in Γ connecting the P_j-edge of $\mathcal{T}$ (or the P_j-edge of Π if the θ-band $\mathcal{T}$ does not exist) with p.

Proof. Let Γ' be the subdiagram of Γ bounded by **top**($\mathcal{T}$) (or by Π if there is no such $\mathcal{T}$), by $\mathcal{B}, \mathcal{B}'$ and p. By Lemma 3.11 and by conditions 2,3, all maximal θ-band of Γ' must start or end on p. Hence the number of such bands is at most $|p|$. There exist at most $|p|$ maximal k-bands in Γ' by Lemma 3.11. Since there exist no (k, θ)-annuli by Lemma 3.11, the number of (k, θ)-cells in Γ' is $O(|p|^2)$. Since $j \neq 1$, the band $\mathcal{T}$ has no $a(P_1)$-cells. Hence every maximal a-band starting on $\mathcal{T}$ must end on a (k, θ)-cell or on p. Then Lemma 6.1 gives the desired upper bound for the length of $\mathcal{T}$.

The maximal P_j-band $\mathcal{B}_1$, that starts on the hub, must end on p and intersect $\mathcal{T}$ (if $\mathcal{T}$ exists). It consists of (P_j, θ)-cells only. Since any two of these cells cannot belong to the same θ-band (by Lemma 3.11), the length of $\mathcal{B}_1$ (or of its part belonging to Γ') is bounded by the number $(O(|p|))$ of maximal θ-bands in Γ'. Hence a subpath of a side of $\mathcal{B}_1$ is the path t that connects $\mathcal{T}$ (or $\partial(\Pi)$) and p and has length $O(|p|)$. □

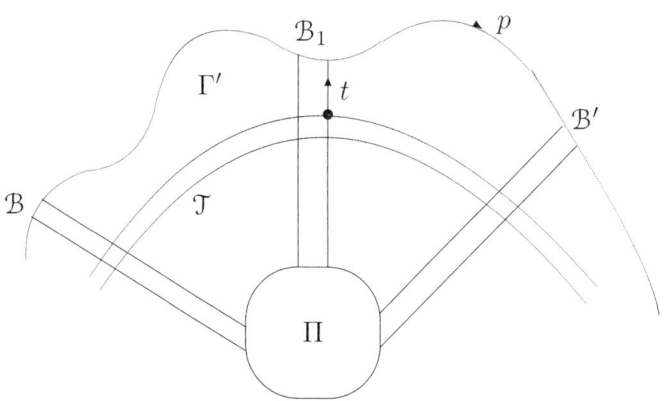

Fig. 35.

Lemma 10.6. *Let V and V' be two words which are conjugate in the group $\mathcal{H}$, and a reduced annular diagram Δ for this conjugacy, contains hubs. Then there is a word U of length $O((|V|+|V'|)^2)$ such that (a) $U = 1$ in $\mathcal{H}$, and a minimal diagram for this equality contains one hub, (b) a minimal diagram for conjugacy of VU and V', or V and $V'U$, has fewer hubs than Δ.*

Proof. Let p and p' be contours of Δ labelled by V and V'. Assume Π is a hub given by Lemmas 10.3 and 10.4. Since $N \geq 8$, the hub is connected with one of contours p, p' by a pair of consecutive K-bands $\mathcal{B}$ and $\mathcal{B}'$ which are not K_1-bands. Without loss of generality, we assume that these two bands ends on p. Besides, there are no other hubs between these k-bands. Denote by Γ the subdiagram bounded by Π, by p and by these two k-bands.

Let $\mathcal{T}$ be the maximal θ-band of Γ that is closest to p. It is of length $O(|p|^2)$ (if $\mathcal{T}$ exists) by Lemma 10.6 and the P_j-edge of $\mathcal{T}$ (or of Π) can be connected with p by a path t of length $O(|p|)$. Lemma 10.1 says that preserving the subdiagram Γ' of Γ which is bounded by $\mathcal{T}$ and p, we can insert a number of non-hub cells (and their mirror images) in Δ, and obtain a diagram $\bar{\Delta}$ (with the same boundary label as Δ) containing a subdiagram Δ' which (1) is attached to Γ' along the top of $\mathcal{T}$; (2) contains exactly one hub; (3) has linearly bounded perimeter as function of $|\mathcal{T}|$, and therefore as function of $|V|$, since $|V| = |p|$.

Then we consider the path t, given by Lemma 10.5, and the boundary path y of Δ'. One may suppose that initial vertices of the paths y and t coincide. Let p_1 be a cyclic permutation of p, starting with the terminal vertex of t. Then the length of the null-homotopic path $t^{-1}yt$ is $O(|p|^2)$, and the path $p_1 t y^{\pm 1} t^{-1}$ bounds an annular diagram with smaller number of hubs than Δ. Thus, to prove the lemma, we just denote the label of $p_0 t y^{\pm 1} t^{-1} p_0^{-1}$ by U where p_0 connects the initial vertices of p and t. □

11 The end of the proof

Lemma 11.1. *The conjugacy problem is decidable for $\mathcal{H}_1$.*

Proof. Let Δ is any annular diagram over $\mathcal{H}_1$. We need to find another diagram with the same boundary labels and recursively bounded number of cells. By Lemma 5.6, we can assume that Δ is reduced and has minimal boundaries.

By Lemma 7.30, we can assume that the boundary of Δ contains θ-edges. Now if $\partial(\Delta)$ contains no k-edges then Δ is a roll and we can refer to Lemma 9.18. Otherwise Δ is a union of a spiral and a recursively bounded number of cells with recursively bounded perimeters (by Lemma 8.4), and we can refer to Lemma 8.6. □

Lemma 11.2. *The conjugacy problem is decidable for $\mathcal{H}$.*

Proof. It suffices to prove that there is an annular diagram over $\mathcal{H}$ for the conjugacy of any two given conjugate words w_1 and w_2, whose number of cells and perimeters of cells are recursively bounded in terms of $|w_1|+|w_2|$. Consider a minimal annular diagram Δ for such a conjugacy. The hub graph $\Gamma(\Delta)$ is a $N-1$ graph by Lemma 10.4. By Lemma 10.3, the number of hubs in Δ does not exceed the number of edges of $\Gamma(\Delta)$, which are incident with the two exterior vertices of $\Gamma(\Delta)$ because $N \geq 8$, that is, the number of hubs is at most the sum of the lengths of boundary paths p and q of Δ.

Then Lemma 10.6 reduces the problem to the conjugacy problem for the group $\mathcal{H}_1$ because the words U from that lemma are conjugates in $\mathcal{H}_1$ of the hub. Thus, the statement follows from Lemma 11.1. □

The next lemma completes the proof of Theorem 1.1

Lemma 11.3. *The embedding of $\mathcal{G}$ into $\mathcal{H}$ is a Frattini embedding.*

Proof. Let u, v be reduced words over $a_1(P_1), ..., a_m(P_1)$ (the generators of the copy of $\mathcal{G}$ in $\mathcal{H}$ by Lemma 3.9). Suppose that u and v are conjugate in $\mathcal{H}$. Consider a minimal conjugacy diagram Δ with boundary labels u and v. Since u and v do not contain k-letters, by Lemmas 10.3 and 10.4, Δ contains no hubs, so Δ is a diagram over $\mathcal{H}_1$.

Without loss of generality we can assume that there are no $\mathcal{G}$-cells which have common edges with the boundary of Δ. Indeed, otherwise we can remove these cells, replacing u (resp. v) by words that are conjugates of u (resp. v) in $\mathcal{G}$.

Suppose that Δ contains k-edges. Then by Lemma 3.11, Δ would contain k-annuli surrounding the hole of Δ. Since the boundary of Δ contains no θ-edges, the k-annuli contain no θ-cells by Lemma 3.11. Therefore its boundary label is a word in $\mathcal{X}$. By Lemma 3.1 the δ-images of u and v are trivial in $\bar{\mathcal{G}}$. Hence they are trivial in $\mathcal{G}$ by Lemma 3.9.

Suppose that Δ contains θ-edges. Then Δ contains a θ-annulus $\mathcal{T}$ surrounding the hole of Δ. Since Δ does not contain k-edges, all θ-edges belong to the same set $\Theta(z)$ or $\bar{\Theta}(z)$. If $z \neq P_1$ then the a-bands starting on a side of $\mathcal{T}$ cannot end on the boundary of Δ or on $\mathcal{G}$-cells. Therefore they must cross $\mathcal{T}$ twice, which contradicts Lemma 3.11. Therefore $z = P_1$. But then all cells in $\mathcal{T}$ are commutativity cells corresponding to relations (2.7). Therefore the labels of the sides of $\mathcal{T}$ coincide and we can remove $\mathcal{T}$ from Δ reducing the number of θ-cells in Δ. This contradicts to the minimality of Δ. Thus Δ contains no θ-edges.

But this means that there are no cells in Δ that have common edges with the contours of Δ. Thus Δ has no cells, and u is a cyclic shift of v (since u, v are reduced). Hence u and v are conjugate in $\mathcal{G}$. □

As we mentioned before, the **proof of Theorem 1.2** proceeds in almost the same way. We need to show that the conjugacy problem in $\mathcal{H}$ is Turing reducible to the conjugacy problem in $\mathcal{G}$. Thus we have to expand the class of recursive functions by the characteristic function of the set of pairs of words (u, v) which are conjugate in $\mathcal{G}$ to the set of elementary recursive functions, and then apply the usual operators used to produce all recursive functions [Mal]. Let us call such functions $\mathcal{G}$-recursive. For example, the word problem in $\mathcal{G}$ is $\mathcal{G}$-recursive, by Clapham's theorem [Cla] the word problem in $\bar{\mathcal{G}}$ is also $\mathcal{G}$-recursive and so on. The reader can check that all the Lemmas in this paper remain true if we replace the word "recursive" in their formulations by "$\mathcal{G}$-recursive". It is easy to see that this modification turns the proof of Theorem 1.1 into a proof of Theorem 1.2.

References

[AC] S. Aanderaa, D.E. Cohen. Modular machines and the Higman-Clapham-Valiev embedding theorem. Word problems, II (Conf. on Decision Problems in Algebra, Oxford, 1976), pp. 17–28, Stud. Logic Foundations Math., 95, North-Holland, Amsterdam-New York, 1980.

[BORS] J. C. Birget, A.Yu. Ol'shanskii, E.Rips, M. V. Sapir. Isoperimetric functions of groups and computational complexity of the word problem. Annals of Mathematics, 156, 2 (2002), 467-518

[Bo] W. W. Boone. Certain simple unsolvable problems in group theory . Proc. Kon. ned. akad. Wetensch. A, (I) 57 (1954), 231–237, (II) 57

(1954), 492–497, (III) 58 (1955), 252–256, (IV) 58 (1955), 571–577, (V) 60 (1957), 22–27, (VI) 60 (1957), 222–232.

[Cla] C. R. J. Clapham. An embedding theorem for finitely generated groups", Proc. London. Math. Soc. (3), 17, 1967, 419-430.

[Col] Donald J. Collins. Conjugacy and the Higman embedding theorem. Word problems, II (Conf. on Decision Problems in Algebra, Oxford, 1976), pp. 81–85, Stud. Logic Foundations Math., 95, North-Holland, Amsterdam-New York, 1980.

[CM] D. J. Collins, C. F. Miller III. The conjugacy problem and subgroups of finite index. Proc. London Math. Soc. (3) 34 (1977), no. 3, 535–556.

[FW] N.J.Fine and H.S.Wilf, Uniqueness theorems for periodic functions, Proc. AMS 16 (1965), 109-114

[GK] A.V. Gorjaga, A.S. Kirkinskiĭ. The decidability of the conjugacy problem cannot be transferred to finite extensions of groups. (Russian) Algebra i Logika 14 (1975), no. 4, 393–406.

[Hi] G. Higman. Subgroups of finitely presented groups. Proc. Roy. Soc. Ser. A, 262 (1961), 455–475.

[Kal] K.A. Kalorkoti. Decision problems in group theory. Proc. London Math. Soc. (3) 44 (1982), no. 2, 312–332.

[KS] O. G. Kharlampovich and M.V. Sapir. Algorithmic problems in varieties. Internat. J. Algebra Comput. 5 (1995), no. 4-5, 379–602.

[KT] *Kourovka Notebook.* Unsolved Problems in Group Theory. 5th edition, Novosibirsk, 1976.

[LS] Roger Lyndon and Paul Schupp. *Combinatorial group theory.* Springer-Verlag, 1977.

[Mak] G. S. Makanin. Equations in a free group. (Russian) Izv. Akad. Nauk SSSR Ser. Mat. 46 (1982), no. 6, 1199–1273, 1344.

[Mal] A.I. Mal'cev. Algorithms and recursive functions. Translated from the first Russian edition by Leo F. Boron, with the collaboration of Luis E. Sanchis, John Stillwell and Kiyoshi Iski Wolters-Noordhoff Publishing, Groningen 1970.

[Ma] Yu. I. Manin. The computable and the non-computable. (Vychislimoe i nevychislimoe). (Russian) [B] Kibernetika. Moskva: "Sovetskoe Radio".

[Mil] Charles F. Miller III. On group-theoretic decision problems and their classification. Annals of Mathematics Studies, No. 68. Princeton University Press, Princeton, N.J.; University of Tokyo Press, Tokyo, 1971.

[Nov] P.S. Novikov, On the algorithmic insolvability of the word problem in group theory. American Mathematical Society Translations, Ser 2, Vol. 9 (1958), pp. 1–122 or Trudy Mat. Inst. im. Steklov. no. 44. Izdat. Akad. Nauk SSSR, Moscow, 1955, 143 pp.

[Ol1] A. Yu. Ol'shanskii. *The geometry of defining relations in groups*, Nauka, Moscow, 1989.

[Ol2] A. Yu. Ol'shanskii. On distortion of subgroups in finitely presented groups. Mat. Sb., 1997, V.188, N 11, 51-98.

[OlSa1] A. Yu. Ol'shanskii, M. V. Sapir. Embeddings of relatively free groups into finitely presented groups. Contemp. Math., 264, 2000, 23-47.

[OlSa2] A.Yu. Ol'shanskii, M. V. Sapir. Length and area functions on groups and quasi-isometric Higman embeddings. Internat. J. Algebra Comput. 11 (2001), no. 2, 137–170.

[OlSa3] A.Yu. Olshanskii, M.V. Sapir. Non-amenable finitely presented torsion-by cyclic groups, Publications of IHES, # 96, 2002.

[OlSa4] A.Yu. Olshanskii, M.V. Sapir. Word, power, order and conjugacy problems in groups. (in preparation)

[Rot] J. Rotman. *An Introduction to the Theory of Groups*. Allyn & Bacon, third edition, 1984

[SBR] M. V. Sapir, J. C. Birget, E. Rips. Isoperimetric and isodiametric functions of groups, Annals of Mathematics, 157, 2 (2002), 345-466.

[Tho] R. J. Thompson. Embeddings into finitely generated simple groups which preserve the word problem. Word problems, II (Conf. on Decision Problems in Algebra, Oxford, 1976), pp. 401–441, Stud. Logic Foundations Math., 95, North-Holland, Amsterdam-New York, 1980.

[Va] M.K.Valiev. On polynomial reducibility of the word problem under embedding of recursively presented groups in finitely presented groups. Mathematical foundations of computer science 1975 (Fourth Sympos., Marinsk Lzně, 1975), pp. 432–438. Lecture Notes in Comput. Sci., Vol. 32, Springer, Berlin, 1975.

Subject index

Accepted word 76
Admissible words of an S-machines 10
Admissible words of the S-machine $\mathcal{S}$ 12
Admissible words of the S-machines $\bar{\mathcal{S}}$ and $\mathcal{S} \cup \bar{\mathcal{S}}$ 17
Age of ω 63
Annular (Schupp) diagram 23
Annulus 30
Auxiliary relations 21
Band 29
k-, θ-, a-bands 30
Base of a θ-band 49
Base of an admissible word 12
Base of a computation 62
Base of a roll 98
Base of a trapezium 50
Basic letters 12
z-best pairing 111
Bottom path of a band 31
Bottom path of a trapezium 50
Boundary of a band 31
Brief history of a computation 63
Cancellation pairing 109
(a,x)-, (k,x)-, (θ,a)-, (θ,k)-cell 27
$\mathcal{G}$-cell 27
Compressible diagram 28
Computation associated with a trapezium 54
$\mathcal{G}$-computation 61
Conditions (R1)-(R4) 45
Connected pair of parentheses 109
Connected pair of letters 110
Contracting (expanding) bands 38
$\bar{\mathcal{E}}$-coordinate 13
Ω-coordinate 13
Data associated with a trapezium 50
a-, k-, θ-, x-edge (letter, cell) 27
Empty band 30
Fractional letters (words) 87
Frattini embedding 2

Free computation 61
Graded presentation 27
z-good pairing 111
l-graph 124
Height of a trapezium 50
Historical period 63
History of a computation 62
History of a roll 98
History of a trapezium (ring) 51
Hub 12
Inner diagram of an annulus 31
Inner part of a sector 12
Inverse rule 10
Locked sectors (by a rule) 13
S-machine 10
Main k-band of a spiral 89
Main relations 20
Maximal band 31
Minimal boundary 48
Minimal diagram 28
Minimal word 48
Minus pair of letters 110
Minus pairing 110
Minus word 110
Normal pair of letters 110
Operation of moving K_j-annuli 121
Operation of removing K_j-annuli 120
$(2,4)$-pairing 111
Plus pair of letters 110
Positive rule 14
$(2,4)$-projection of a word 110
Quasiring 51
Quasispiral 89
Quasitrapezium 51
Rank of a cell 27
Reduced computation 59
Reduced or non-reduced diagram 28
Reducible pair of cells 28
τ-regular word 48
Related uniform words 37
(a,x)-, (k,x)-, (θ,a)-, (θ,k)-relations 21
$\mathcal{G}$-relations 27
Ring 50

Ring computation 62
Roll 97
S-rule 10
S-rule active with respect to a sector 19
S-rule applicable to a word 14
Sector of an admissible word 12
a-similar rules 110
Small trapezium 56
Spiral 86
Standard computation 64
State letters 10
Tame computation 73
Tape letters 12
Top path of a band 31
Top path of a trapezium 50
Transition rules 15
Trapezium 49
Trapezium of the first (second, mixed) type 54
Turn of a spiral 94
Type of a diagram 27
Type of a path 45
Uniform word 37
van Kampen diagram 22
Wild computation 73

$\xrightarrow{\ell}$ 13
$\equiv$ 12
$\mathcal{A}$ 12
$\mathcal{A}(z)$ 12
$\bar{\mathcal{A}}$ 17
α_τ 20
β 19
bot(.) 31
br(.) 63
δ 23
diff(.) 65
$\mathcal{E}$ 3
$\bar{\mathcal{E}}$ 11
$\mathcal{G}$ 11
$\bar{\mathcal{G}}$ 11

γ 72
$\mathcal{H}$ 22
$\mathcal{H}_0$ 26
$\mathcal{H}_1$ 27
$\mathcal{H}_2$ 34
$\mathcal{K}$ 11
$\tilde{\mathcal{K}}$ 12
$\bar{\mathcal{K}}$ 17
$\tilde{\bar{\mathcal{K}}}$ 17
$\overleftarrow{L}_j$ 12
$\overrightarrow{R}_j$ 12
Σ 12
$\tilde{\Sigma}$ 12
$\Sigma(w)$ 16
$\Sigma(w_1, w_2, w_3, w_4)$ 24
$\bar{\Sigma}(w_1, w_2, w_3, w_4)$ 24
$\Sigma_{r,i}(w_1, w_2, w_3, w_4)$ 24
$\bar{\Sigma}_{r,i}(w_1, w_2, w_3, w_4)$ 24
$\mathcal{S}(\omega)$ 14
Θ 20
$\Theta(\tau)$ 20
$\Theta(z)$ 20
$\bar{\Theta}$ 22
top(.) 31
$U(r,i)$ 12
$\mathcal{X}$ 20
$\mathcal{X}(\tau)$ 20
$\mathcal{X}(z)$ 20
$\mathcal{X}(z,\tau)$ 20
z_- 12
z_+ 12

Alexander Yu. Ol'shanskii
Department of Mathematics
Vanderbilt University
alexander.olshanskiy@vanderbilt.edu
and
Department of Higher Algebra
MEHMAT
Moscow State University
olshan@shabol.math.msu.su

Mark V. Sapir
Department of Mathematics
Vanderbilt University
msapir@math.vanderbilt.edu

Editorial Information

To be published in the *Memoirs*, a paper must be correct, new, nontrivial, and significant. Further, it must be well written and of interest to a substantial number of mathematicians. Piecemeal results, such as an inconclusive step toward an unproved major theorem or a minor variation on a known result, are in general not acceptable for publication. Papers appearing in *Memoirs* are generally longer than those appearing in *Transactions*, which shares the same editorial committee.

As of March 1, 2004, the backlog for this journal was approximately 4 volumes. This estimate is the result of dividing the number of manuscripts for this journal in the Providence office that have not yet gone to the printer on the above date by the average number of monographs per volume over the previous twelve months, reduced by the number of volumes published in four months (the time necessary for preparing a volume for the printer). (There are 6 volumes per year, each containing at least 4 numbers.)

A Consent to Publish and Copyright Agreement is required before a paper will be published in the *Memoirs*. After a paper is accepted for publication, the Providence office will send a Consent to Publish and Copyright Agreement to all authors of the paper. By submitting a paper to the *Memoirs*, authors certify that the results have not been submitted to nor are they under consideration for publication by another journal, conference proceedings, or similar publication.

Information for Authors

Memoirs are printed from camera copy fully prepared by the author. This means that the finished book will look exactly like the copy submitted.

The paper must contain a *descriptive title* and an *abstract* that summarizes the article in language suitable for workers in the general field (algebra, analysis, etc.). The *descriptive title* should be short, but informative; useless or vague phrases such as "some remarks about" or "concerning" should be avoided. The *abstract* should be at least one complete sentence, and at most 300 words. Included with the footnotes to the paper should be the 2000 *Mathematics Subject Classification* representing the primary and secondary subjects of the article. The classifications are accessible from www.ams.org/msc/. The list of classifications is also available in print starting with the 1999 annual index of *Mathematical Reviews*. The Mathematics Subject Classification footnote may be followed by a list of *key words and phrases* describing the subject matter of the article and taken from it. Journal abbreviations used in bibliographies are listed in the latest *Mathematical Reviews* annual index. The series abbreviations are also accessible from www.ams.org/publications/. To help in preparing and verifying references, the AMS offers MR Lookup, a Reference Tool for Linking, at www.ams.org/mrlookup/. When the manuscript is submitted, authors should supply the editor with electronic addresses if available. These will be printed after the postal address at the end of the article.

Electronically prepared manuscripts. The AMS encourages electronically prepared manuscripts, with a strong preference for $\mathcal{A}_{\mathcal{M}}\mathcal{S}$-LaTeX. To this end, the Society has prepared $\mathcal{A}_{\mathcal{M}}\mathcal{S}$-LaTeX author packages for each AMS publication. Author packages include instructions for preparing electronic manuscripts, the *AMS Author Handbook*, samples, and a style file that generates the particular design specifications of that publication series. Though $\mathcal{A}_{\mathcal{M}}\mathcal{S}$-LaTeX is the highly preferred format of TeX, author packages are also available in $\mathcal{A}_{\mathcal{M}}\mathcal{S}$-TeX.

Authors may retrieve an author package from e-MATH starting from `www.ams.org/tex/` or via FTP to `ftp.ams.org` (login as `anonymous`, enter username as password, and type `cd pub/author-info`). The *AMS Author Handbook* and the *Instruction Manual* are available in PDF format following the author packages link from `www.ams.org/tex/`. The author package can be obtained free of charge by sending email to `pub@ams.org` (Internet) or from the Publication Division, American Mathematical Society, 201 Charles St., Providence, RI 02904, USA. When requesting an author package, please specify $\mathcal{AMS}$-LaTeX or $\mathcal{AMS}$-TeX, Macintosh or IBM (3.5) format, and the publication in which your paper will appear. Please be sure to include your complete mailing address.

Sending electronic files. After acceptance, the source file(s) should be sent to the Providence office (this includes any TeX source file, any graphics files, and the DVI or PostScript file).

Before sending the source file, be sure you have proofread your paper carefully. The files you send must be the EXACT files used to generate the proof copy that was accepted for publication. For all publications, authors are required to send a printed copy of their paper, which exactly matches the copy approved for publication, along with any graphics that will appear in the paper.

TeX files may be submitted by email, FTP, or on diskette. The DVI file(s) and PostScript files should be submitted only by FTP or on diskette unless they are encoded properly to submit through email. (DVI files are binary and PostScript files tend to be very large.)

Electronically prepared manuscripts can be sent via email to `pub-submit@ams.org` (Internet). The subject line of the message should include the publication code to identify it as a Memoir. TeX source files, DVI files, and PostScript files can be transferred over the Internet by FTP to the Internet node `e-math.ams.org` (130.44.1.100).

Electronic graphics. Comprehensive instructions on preparing graphics are available at `www.ams.org/jourhtml/graphics.html`. A few of the major requirements are given here.

Submit files for graphics as EPS (Encapsulated PostScript) files. This includes graphics originated via a graphics application as well as scanned photographs or other computer-generated images. If this is not possible, TIFF files are acceptable as long as they can be opened in Adobe Photoshop or Illustrator. No matter what method was used to produce the graphic, it is necessary to provide a paper copy to the AMS.

Authors using graphics packages for the creation of electronic art should also avoid the use of any lines thinner than 0.5 points in width. Many graphics packages allow the user to specify a "hairline" for a very thin line. Hairlines often look acceptable when proofed on a typical laser printer. However, when produced on a high-resolution laser imagesetter, hairlines become nearly invisible and will be lost entirely in the final printing process.

Screens should be set to values between 15% and 85%. Screens which fall outside of this range are too light or too dark to print correctly. Variations of screens within a graphic should be no less than 10%.

Inquiries. Any inquiries concerning a paper that has been accepted for publication should be sent directly to the Electronic Prepress Department, American Mathematical Society, 201 Charles St., Providence, RI 02904, USA.

Editors

This journal is designed particularly for long research papers, normally at least 80 pages in length, and groups of cognate papers in pure and applied mathematics. Papers intended for publication in the *Memoirs* should be addressed to one of the following editors. In principle the Memoirs welcomes electronic submissions, and some of the editors, those whose names appear below with an asterisk (*), have indicated that they prefer them. However, editors reserve the right to request hard copies after papers have been submitted electronically. Authors are advised to make preliminary email inquiries to editors about whether they are likely to be able to handle submissions in a particular electronic form.

*Algebra to ROBERT GURALNICK, Department of Mathematics, University of Southern California, Los Angeles, CA 90089-1113; email: guralnic@math.usc.edu

Algebraic geometry to DAN ABRAMOVICH, Department of Mathematics, Boston University, 111 Cummington St., Boston, MA 02215; email: abramovic@bu.edu

*Algebraic number theory to V. KUMAR MURTY, Department of Mathematics, University of Toronto, 100 St. George Street, Toronto, ON M5S 1A1, Canada; email: murty@math.toronto.edu

Combinatorics and Lie theory to SERGEY FOMIN, Department of Mathematics, University of Michigan, Ann Arbor, Michigan 48109-1109; email: fomin@umich.edu

Complex analysis and complex geometry to DUONG H. PHONG, Department of Mathematics, Columbia University, 2990 Broadway, New York, NY 10027-0029; email: phong@math.columbia.edu

*Differential geometry and global analysis to LISA C. JEFFREY, Department of Mathematics, University of Toronto, 100 St. George St., Toronto, ON Canada M5S 3G3; email: jeffrey@math.toronto.edu

Dynamical systems and ergodic theory to ROBERT F. WILLIAMS, Department of Mathematics, University of Texas, Austin, Texas 78712-1082; email: bob@math.utexas.edu

*Functional analysis and operator algebras to MARIUS DADARLAT, Department of Mathematics, Purdue University, 150 N. University St., West Lafayette, IN 47907-2067; email: mdd@math.purdue.edu

*Geometric analysis to TOBIAS COLDING, Courant Institute, New York University, 251 Mercer St., New York, NY 10012; email: colding@cims.nyu.edu

*Geometric analysis to MLADEN BESTVINA, Department of Mathematics, University of Utah, 155 South 1400 East, JWB 233, Salt Lake City, Utah 84112-0090; email: bestvina@math.utah.edu

Harmonic analysis to ALEXANDER NAGEL, Department of Mathematics, University of Wisconsin, 480 Lincoln Drive, Madison, WI 53706-1313; email: nagel@math.wisc.edu

Harmonic analysis, representation theory, and Lie theory to ROBERT J. STANTON, Department of Mathematics, The Ohio State University, 231 West 18th Avenue, Columbus, OH 43210-1174; email: stanton@math.ohio-state.edu

*Logic to STEFFEN LEMPP, Department of Mathematics, University of Wisconsin, 480 Lincoln Drive, Madison, Wisconsin 53706-1388; email: lempp@math.wisc.edu

Number theory to HAROLD G. DIAMOND, Department of Mathematics, University of Illinois, 1409 W. Green St., Urbana, IL 61801-2917; email: diamond@math.uiuc.edu

*Ordinary differential equations, and applied mathematics to PETER W. BATES, Department of Mathematics, Michigan State University, East Lansing, MI 48824-1027; email: peter@math.msu.edu

*Partial differential equations to PATRICIA E. BAUMAN, Department of Mathematics, Purdue University, West Lafayette, IN 47907-1395; email: bauman@math.purdue.edu

*Probability and statistics to KRZYSZTOF BURDZY, Department of Mathematics, University of Washington, Box 354350, Seattle, Washington 98195-4350; email: burdzy@math.washington.edu

*Real analysis and partial differential equations to DANIEL TATARU, Department of Mathematics, University of California, Berkeley, Berkeley, CA 94720; email: tataru@ math.berkeley.edu

All other communications to the editors should be addressed to the Managing Editor, WILLIAM BECKNER, Department of Mathematics, University of Texas, Austin, TX 78712-1082; email: beckner@math.utexas.edu.

Titles in This Series

807 **Carlos A. Cabrelli, Christopher Heil, and Ursula M. Molter,** Self-similarity and multiwavelets in higher dimensions, 2004

806 **Spiros A. Argyros and Andreas Tolias,** Methods in the theory of hereditarily indecomposable Banach spaces, 2004

805 **Philip L. Bowers and Kenneth Stephenson,** Uniformizing dessins and Belyĭ maps via circle packing, 2004

804 **A. Yu. Ol'shanskii and M. V. Sapir,** The conjugacy problem and Higman embeddings, 2004

803 **Michael Field and Matthew Nicol,** Ergodic theory of equivariant diffeomorphisms: Markov partitions and stable ergodicity, 2004

802 **Martin W. Liebeck and Gary M. Seitz,** The maximal subgroups of positive dimension in exceptional algebraic groups, 2004

801 **Fabio Ancona and Andrea Marson,** Well-posedness for general 2×2 systems of conservation laws, 2004

800 **V. Poénaru and C. Tanasi,** Equivariant, almost-arborescent representations of open simply-connected 3-manifolds; A finiteness result, 2004

799 **Barry Mazur and Karl Rubin,** Kolyvagin systems, 2004

798 **Benoît Mselati,** Classification and probabilistic representation of the positive solutions of a semilinear elliptic equation, 2004

797 **Ola Bratteli, Palle E. T. Jorgensen, and Vasyl' Ostrovs'kyĭ,** Representation theory and numerical AF-invariants, 2004

796 **Marc A. Rieffel,** Gromov-Hausdorff distance for quantum metric spaces/Matrix algebras converge to the sphere for quantum Gromov-Hausdorff distance, 2004

795 **Adam Nyman,** Points on quantum projectivizations, 2004

794 **Kevin K. Ferland and L. Gaunce Lewis, Jr.,** The $RO(G)$-graded equivariant ordinary homology of G-cell complexes with even-dimensional cells for $G = \mathbb{Z}/p$, 2004

793 **Jindřich Zapletal,** Descriptive set theory and definable forcing, 2004

792 **Inmaculada Baldomá and Ernest Fontich,** Exponentially small splitting of invariant manifolds of parabolic points, 2004

791 **Eva A. Gallardo-Gutiérrez and Alfonso Montes-Rodríguez,** The role of the spectrum in the cyclic behavior of composition operators, 2004

790 **Thierry Lévy,** Yang-Mills measure on compact surfaces, 2003

789 **Helge Glöckner,** Positive definite functions on infinite-dimensional convex cones, 2003

788 **Robert Denk, Matthias Hieber, and Jan Prüss,** $\mathcal{R}$-boundedness, Fourier multipliers and problems of elliptic and parabolic type, 2003

787 **Michael Cwikel, Per G. Nilsson, and Gideon Schechtman,** Interpolation of weighted Banach lattices/A characterization of relatively decomposable Banach lattices, 2003

786 **Arnd Scheel,** Radially symmetric patterns of reaction-diffusion systems, 2003

785 **R. R. Bruner and J. P. C. Greenlees,** The connective K-theory of finite groups, 2003

784 **Desmond Sheiham,** Invariants of boundary link cobordism, 2003

783 **Ethan Akin, Mike Hurley, and Judy A. Kennedy,** Dynamics of topologically generic homeomorphisms, 2003

782 **Masaaki Furusawa and Joseph A. Shalika,** On central critical values of the degree four L-functions for GSp(4): The Fundamental Lemma, 2003

781 **Marcin Bownik,** Anisotropic Hardy spaces and wavelets, 2003

780 **S. Marmi and D. Sauzin,** Quasianalytic monogenic solutions of a cohomological equation, 2003

779 **Hansjörg Geiges,** h-principles and flexibility in geometry, 2003

TITLES IN THIS SERIES

778 **David B. Massey**, Numerical control over complex analytic singularities, 2003
777 **Robert Lauter**, Pseudodifferential analysis on conformally compact spaces, 2003
776 **U. Haagerup, H. P. Rosenthal, and F. A. Sukochev**, Banach embedding properties of non-commutative L^p-spaces, 2003
775 **P. Lochak, J.-P. Marco, and D. Sauzin**, On the splitting of invariant manifolds in multidimensional near-integrable Hamiltonian systems, 2003
774 **Kai A. Behrend**, Derived ℓ-adic categories for algebraic stacks, 2003
773 **Robert M. Guralnick, Peter Müller, and Jan Saxl**, The rational function analogue of a question of Schur and exceptionality of permutation representations, 2003
772 **Katrina Barron**, The moduli space of $N=1$ superspheres with tubes and the sewing operation, 2003
771 **Shigenori Matsumoto**, Affine flows on 3-manifolds, 2003
770 **W. N. Everitt and L. Markus**, Elliptic partial differential operators and symplectic algebra, 2003
769 **Jie Wu**, Homotopy theory of the suspensions of the projective plane, 2003
768 **R. Höpfner and E. Löcherbach**, Limit theorems for null recurrent Markov processes, 2003
767 **Po Hu**, S-modules in the category of schemes, 2003
766 **Su Gao and Alexander S. Kechris**, On the classification of Polish metric spaces up to isometry, 2003
765 **Robert Bieri and Ross Geoghegan**, Connectivity properties of group actions on non-positively curved spaces, 2003
764 **J. Spandaw**, Noether-Lefschetz problems for degeneracy loci, 2003
763 **Yasuyuki Kachi and Eiichi Sato**, Segre's reflexivity and an inductive characterization os hyperquadrics, 2002
762 **Leiba Rodman, Ilya M. Spitkovsky, and Hugo Woerdeman**, Abstract band method via factorization, positive and band extensions of multivariable almost periodic matrix functions, and spectral estimation, 2002
761 **Oliver Druet and Emmanuel Hebey**, The AB program in geometric analysis : Sharp Sobolev inequalities and related problems, 2002
760 **Markus Banagl**, Extending intersection homology type invariants to non-Witt spaces, 2002
759 **Donald M. Davis**, From representation theory to homotopy groups, 2002
758 **Alan Forrest, John Hunton, and Johannes Kellendonk**, Topological invariants for projection method patterns, 2002
757 **Douglas Bowman**, q-difference operators, orthogonal polynomials, and symmetric expansions, 2002
756 **José Ignacio Cogolludo-Agustín**, Topological invariants of the complement to arrangements of rational plane curves, 2002
755 **M. A. Mandell and J. P. May**, Equivariant orthogonal spectra and S-modules, 2002
754 **Edward L. Green, Idun Reiten, and Øyvind Solberg**, Dualities on generalized Koszul algebras, 2002
753 **Daniel Panazzolo**, Desingularization of nilpotent singularities in families of planar vector fields, 2002
752 **Linus Kramer**, Homogeneous spaces, Tits buildings, and isoparametric hypersurfaces, 2002

For a complete list of titles in this series, visit the
AMS Bookstore at **www.ams.org/bookstore/**.